THE SURFACE PROPERTIES OF OXIDIZED SILICON

THE SURFACE PROPERTIES OF OXIDIZED SILICON

E. KOOI

Springer-Verlag Berlin Heidelberg GmbH

This book was written as a thesis for a doctor's degree in applied physics at the Technical University, Eindhoven.

This book contains x + 134 pages and 57 illustrations.

U.D.C. No. 621.382 : 539.232

Library of Congress Catalog Card Number: 67-28721

ISBN 978-3-662-39204-1 ISBN 978-3-662-40210-8 (eBook)
DOI 10.1007/978-3-662-40210-8

Originally published by N.V. Philips' Gloeilampenfabrieken, Eindhoven in 1967
Softcover reprint of the hardcover 1st edition 1967

PREFACE

Understanding of the surface properties of oxidized silicon is of major importance in the modern semiconductor device technology. The investigations described in this work were done in the Philips Research Laboratories in Eindhoven and carried out in order to find relationships between the preparation method of the oxidized surface and the resulting device properties. The book originally appeared in thesis form and a part of it was published earlier in a number of separate articles.

Two chapters are devoted to effects of ionizing irradiations on oxidized silicon. Such effects appeared suitable to obtain more information about the oxide and oxide-silicon interface structure. The observed phenomena explain why MOS transistors and other silicon devices can be sensitive to ionizing irradiations and how this sensitivity depends on the oxide preparation method.

The author greatfully acknowledges M.V. Whelan for a pleasant cooperation and Prof. L. J. Tummers for his stimulation to present the work as a thesis.

E. Kooi

June 1967

CONTENTS

1. INTRODUCTION

Semiconductor surfaces have been investigated for many years. Since the discovery of the transistor effect (1948), these investigations have been concentrated mainly on germanium and silicon crystals, particularly because various surface phenomena were found to have a detrimental influence on the properties and the stability of *pn*-junction devices made with these materials. In 1959 it was reported by Atalla and coworkers [1-1]) that thermally grown silicon-dioxide films can have a stabilizing action on the surface properties of silicon. Since about that time SiO_2 coatings have found a wide-spread use in the silicon-transistor technology, although not only for the reason of surface stabilization. The films are also employed for selective masking against impurity diffusions, used to make *pn* junctions in the silicon. They can also separate metal electrodes from the semiconductor. Such metal electrodes may be used for example to connect elements such as transistors, diodes and resistors made in silicon "integrated circuits".

The work to be described in this book was carried out to obtain a better understanding of various functions of the oxide film and the properties of the boundary between the film and the silicon substrate. Chapter 2 is a general discussion of surface phenomena in semiconducting material and possibilities of SiO_2 coatings. The first sections (2.1-2.4) review a number of possible effects of impurities and crystal defects on the bulk and surface properties of silicon. After that various technological aspects of oxide coatings are considered (sec. 2.5). One of these aspects is the possibility to construct MOS (metal-oxide-semiconductor) transistors. The action of these devices is based on modifying the surface conductance of a semiconductor by varying a voltage applied across the MOS structure. Their properties are therefore strongly dependent on the surface properties of the semiconductor. In the work to be reported they were often used as tools to study surface phenomena in oxidized silicon. The construction of MOS transistors and the influence of surface imperfections on their properties are therefore reviewed in some more detail (sec. 2.6). Another method which was used to obtain information on the surface properties of oxidized silicon was measurement of the differential capacitance of a MOS structure as a function of a d.c. voltage applied across it. This method is described in sec. 2.7. Several oxide-preparation methods and properties of SiO_2 films are given in secs 2.8 and 2.9. The last two sections of chapter 2 give a survey of the influence of oxidation and further treatments on the resulting surface properties (sec. 2.10) and of various stability problems which may be encountered in oxide-covered silicon devices (sec. 2.11).

The chapters 3 to 7 give detailed descriptions of studies carried out by the author. These chapters are (nearly) equal to a number of papers [1-2-6]), published elsewhere. (A part of the work has also been described in a review paper [1-7]).

Chapter 3 considers the formation of surface films during diffusion of phosphorus into silicon and the masking action of SiO_2 films against such diffusions. The masking action will be shown to be accompanied by the formation of a mixed oxide of silicon and phosphorus on the top of the SiO_2 films. In the following chapters it will appear that the presence of such a "phosphate-glass" layer can be of large influence on various phenomena, observed in oxidized silicon.

These phenomena are related to heat treatments (chapters 4 and 7) and ionizing irradiations (chapters 5 and 6). To explain the results it will appear to be necessary to take into account the influence of imperfections at the Si-SiO_2 interface as well as the possibility of charge being distributed through the oxide film. Both the number of imperfections and the charge distribution may change during heat treatments and ionizing irradiations. In practice this also means that certain properties of oxide-covered devices may change during such treatments. Such instability effects are generally increased when electric-potential differences are present in the oxidized-silicon structure. These effects are tried to be understood in order to find ways to improve the properties of oxidized silicon.

A large number of heat-treatment and irradiation effects show that the presence of hydrogen can have a large influence on the oxide and interface structure. In chapter 7 it will be shown that alkali (sodium) impurities can play an important role too, although in the model of oxidized silicon given there a role of hydrogen is again obvious. Finally it will appear to be possible to make fairly stable and ideal oxidized silicon surfaces, i.e. surfaces with a very small number of effective interface and oxide imperfections.

REFERENCES

[1-1]) M. M. Atalla, E. Tannenbaum and E. J. Scheibner, Bell Sys. techn. J. **38**, 749-783, 1959.
[1-2]) E. Kooi, J. electrochem. Soc. **111**, 1383-1387, 1964 (nearly equal to chapter 3 of this book).
[1-3]) ——— , Phil. Res. Repts **20**. 578-594, 1965 (chapter 4 of this book).
[1-4]) ——— , Phil. Res. Repts **20**, 306-314, 1965 (chapter 5 of this book).
[1-5]) ——— , Phil. Res. Repts **20**, 595-619, 1965 (chapter 6 of this book).
[1-6]) ——— , Phil. Res. Repts **21**, 477-495, 1966 (chapter 7 of this book).
[1-7]) ——— , IEEE Trans. **ED-13**, 238-245, 1966.

2. GENERAL REVIEW OF THE EFFECT OF SILICON-DIOXIDE COATINGS ON THE SURFACE PROPERTIES OF SILICON AND THE IMPORTANCE OF THESE COATINGS IN SEMICONDUCTOR-DEVICE TECHNOLOGY

2.1. Bulk properties of silicon

At room temperature pure (intrinsic) silicon does not contain many mobile charge carriers: about $1{\cdot}6.10^{10}$ electrons and the same number of positively charged holes per cm^3. Impurities can be incorporated to get an increased number of either electrons or holes. Suitable donor elements are found in the fifth column of the periodic system, for example P, As, Sb. When atoms of these elements are built in substitutionally in the silicon lattice, their ionization energy is relatively low. At room temperature the equilibrium of the reaction (D = donor impurity, e^- = electron)

$$D \rightleftarrows D^+ + e^- \tag{2.1}$$

lies almost completely to the right-hand side. In the energy-band picture of silicon this type of donors can be represented by donor states close to the conduction band (fig. 2.1).

Other impurities, like those of the third column of the periodic system (B, Al, Ga, In) tend to accept an electron (give off a hole) when they are incorporated in the silicon lattice. This acceptor action can be written as

$$A + e^- \rightleftarrows A^- \quad \text{or} \quad A \rightleftarrows A^- + e^+, \tag{2.2}$$

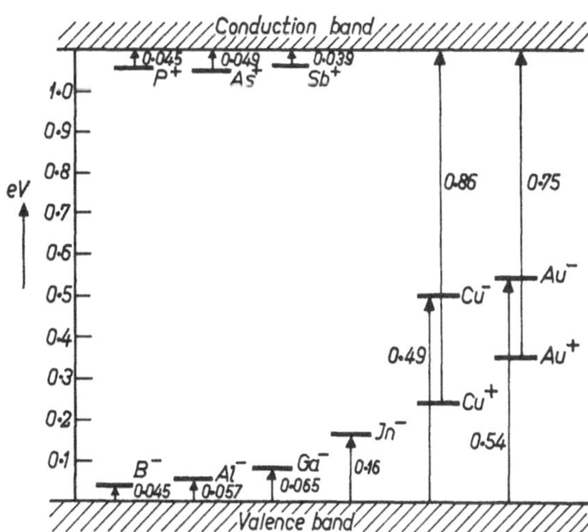

Fig. 2.1. Energy levels of a number of donor and acceptor impurities indicated in the energy gap of silicon (1·1 eV at 300 °K). A more complete review can be found in ref. 2-1, fig. 8, 12.

where A is the acceptor impurity and e^+ is a positive hole. In the energy-band scheme of silicon these acceptors give rise to energy states close to the valence band. Other impurities like Au and Cu have much higher ionization energies (fig. 2.1). Their effect on the electrical properties of silicon is rather intricate. Taking Au as an example, it can be seen that this element can act as a donor as well as an acceptor, which can be indicated by the following equilibria:

$$\text{donor action} \quad : \text{Au} \rightleftarrows \text{Au}^+ + e^-, \tag{2.3}$$

$$\text{acceptor action}: \text{Au} \rightleftarrows \text{Au}^- + e^+. \tag{2.4}$$

The consequence is that gold tends to capture electrons in n-type and holes in p-type silicon. Consequently both types of material become nearly intrinsic (i.e. electron and hole concentration equal) when a sufficient amount of gold is incorporated.

Gold is introduced in many silicon devices in order to modify the lifetime of electrons and holes in the material. The presence of gold centres makes recombination of electrons and holes easier. Incorporation of such recombination centres is of importance in cases where storage of holes and electrons has to be limited, for example in devices to be used for fast-switching purposes.

In fig. 2.1 no energy levels have been indicated for lattice defects like silicon vacancies and interstitials. It seems that their influence on the electron and hole concentrations is relatively small in pure single-crystal silicon. After quick cooling of a silicon sample the average electron and hole lifetimes appear in general to have decreased. This may be due, in any case partly, to the presence of silicon vacancies and interstitials. Dislocations are often observed in single-crystal silicon, but their effect on the average electron and hole concentration in the crystal is small.

2.2. Imperfections at the surface of a silicon crystal

The ending of the regular lattice at the surface of a crystal is an obvious reason for the presence of centres which may influence the semiconducting properties of the silicon. At a "clean" silicon surface, i.e. a surface at which there are no foreign elements present, unsaturated (perhaps "dangling") silicon bonds may occur. A "clean" silicon surface may be made and maintained in a high-vacuum system. In practice, however, a silicon surface is always covered by some oxide film. The structure of this film, the nature of its boundary to the silicon, and the possible presence of various impurities can then be assumed to have a considerable effect on the semiconducting properties of the silicon near the surface.

The various possible centres at the semiconductor surface may be distinguished in the same types as those occurring in the bulk of the crystal. Donor- as well as acceptor-type centres may be present, which according to reaction (2.1) resp. (2.2) tend to make a region near the surface n-type or p-type. Whether this happens or not depends on the concentration and the ionization energies of these imperfections and the impurity concentrations in the crystal. It is common usage to refer to the energy levels due to surface centres by the term "surface states". Surface states in the energy gap of the silicon have been indicated in fig. 2.2 as if they are lying in a two-dimensional interface between the oxide and the semiconductor. In most cases this has to be considered as an approximation, as the states may also occur in the oxide or in the silicon at some distance from the interface. That thermal oxidation of silicon may induce considerable deviations in the concentration of donor and acceptor elements in the surface region from that in the bulk will be discussed later. It is further questionable in how far the junction between a thermally grown oxide film and the silicon substrate can be considered as a two-dimensional plane. The junction between the two phases may be gradual and it is hard to say how thick the interface region is.

Donor-type surface states occurring not too close to the conduction band and acceptor states situated not too close to the valence band will tend to trap excess holes resp. electrons near the surface and tend thus to make a silicon region near the surface high-ohmic. These effects are reminiscent of the effect of gold in bulk material.

Defects or impurities at the surface may also give rise to an increased recombination and generation velocity of holes and electrons. Surface recombination effects can affect the characteristics of *pn*-junction diodes and influence the amplification processes in transistors. A profound discussion of these effects falls outside the scope of this work.

2.3. Physical model of an oxide-coated semiconductor surface

2.3.1. *The surface potential*

The presence of donors and/or acceptors due to imperfections at the surface can result in a difference in electron and hole concentrations between the surface and the bulk. Electronic equilibrium between surface and bulk can then only exist if there is a potential difference which prevents the electrons and holes from diffusing from the places of high concentration to the regions of low concentrations (it may also be said that this potential difference is set up because of some displacement of electrons and holes from sites with high concentration to regions with low concentration). In oxidized silicon often a sheet of positive charge seems to be present in the oxide near the interface; this oxide charge induces negative charge at the silicon surface. The sheet of oxide charge and interface states have been indicated in fig. 2.2.

In this figure it is further shown how the differences between surface and bulk material can be indicated in an energy scheme. In this scheme the energy gap (1·1 eV for silicon at room temperature) between the valence and conduction

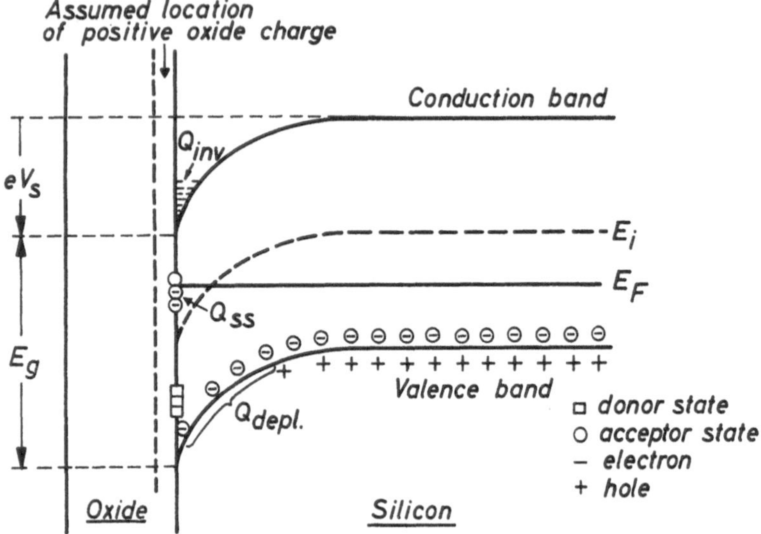

Fig. 2.2. Model of an oxide-coated semiconductor surface. The given charge- and surface (interface)-state distribution may be considered to be representative for many thermally oxidized p-type silicon samples. The negative space charge in the silicon, caused by the presence of positively charged oxide centres, can be divided into a depletion charge Q_{depl} (acceptor centres in a region depleted of holes) and an inversion charge Q_{inv} (free electrons in a layer close to the surface). The presence of a space charge is accompanied by a band bending in the negative direction and a positive surface potential V_s. Note that the band bending gives donor-type surface states a tendency to be neutral and acceptor states a tendency to be charged negatively. In this way the charge Q_{ss} in the interface states counteracts the effect of oxide charge on the surface potential.

band has been shown. The equilibrium density of electrons and holes is given by the position of the Fermi level E_{F} with respect to the middle E_i of the energy gap between the bands:

$$n = n_i \exp{(E_{\mathrm{F}} - E_i)/kT} \tag{2.5}$$

and

$$p = n_i \exp{(E_i - E_{\mathrm{F}})/kT}, \tag{2.6}$$

where n_i is equal to the number of electrons or holes in intrinsic material, i.e. when E_{F} coincides with E_i, k is Boltzmann's constant and T the temperature in degrees Kelvin.

In the example given in fig. 2.2 the material is p-type, indicated by the Fermi level being below the middle of the energy gap and caused by the presence of acceptor levels. Near the surface the bands bend in such a way that the material becomes there n-type. The position of the bands at the surface with respect to the bulk, the surface barrier height or band bending E_s gives also the potential difference V_s between surface and bulk, often indicated as surface potential. In the case of fig. 2.2 V_s is positive ($E_s = -eV_s$; e is the positive unit charge: $1{\cdot}6.10^{-19}$ coulomb).

Depending on the band bending and the type of material various surface conditions may be distinguished. In the case of fig. 2.2 an n-type *inversion* layer is present at the surface of the p-type crystal. If the band bending would have been less but of the same sign it might have been insufficient for inversion. Then the surface layer would have been less p-type than the bulk. Such a layer may be indicated as a *depletion* layer. *Accumulation* of holes near the surface will occur when the band bending is in the opposite direction (V_s negative).

In the case where the bulk material is n-type, accumulation of electrons corresponds to a positive value of V_s whereas depletion of electrons and the formation of a p-type inversion layer may occur when V_s is negative.

The sign and the value of the surface potential depend on the number and the properties of the various surface centres present. A band bending and a distribution of surface states like that shown in fig. 2.2 is often found in thermally oxidized silicon. It may be noted that the interface states fulfil here the function of traps. The charge in the acceptor-type states counteracts the effect of the oxide charge.

2.3.2. *The field effect; fast and slow surface states*

It is often possible to change the surface potential by applying a voltage between the semiconductor and a field plate opposite to its surface. The changes in band bending may be noted by observing changes in the contribution of the surface region to the conductance of the semiconductor sample. The principle of these field-effect measurements and the information which they may give will now be considered.

The conductance G of a sheet of semiconductor material is given by its dimensions, the concentrations of electrons (n) and holes (p) and their mobility (μ_n and μ_p). In the case of a rectangular sheet with contacts on two parallel planes (fig. 2.3) it is

$$G = \frac{abe}{l}(n\mu_n + p\mu_p); \tag{2.7}$$

a, b and l are indicated in fig. 2.3, e is the unit charge ($1{\cdot}6.10^{-19}$ coulomb).

In eq. (2.7) the conductivity has been assumed to be the same everywhere in the crystal. But in the previous section we have seen that the concentration of holes and electrons at the surface may differ from those in the interior of the crystal. An increase in the contribution to the conductance is expected when an accumulation or strong inversion layer is present near the surface and a reduction when the region near the surface is depleted of charge carriers. Such layers may be induced by applying a voltage between a field plate and the semiconductor (fig. 2.3), and one may calculate how the surface conductance is expected to depend on the applied voltage. Many measurements of this kind have been published [2-2]), especially for germanium surfaces. As only the con-

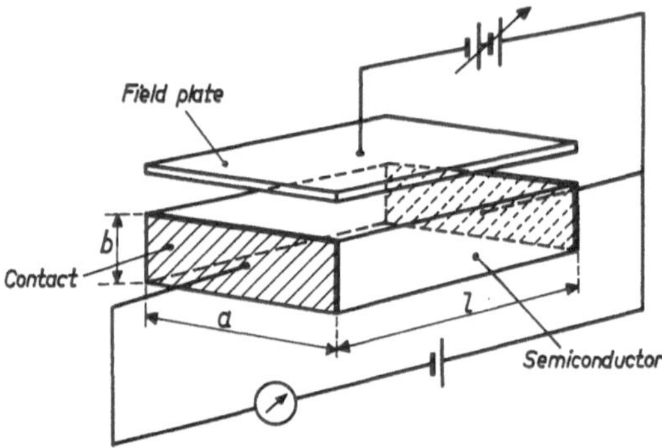

Fig. 2.3. Field-effect experiment in which the conductance of a sheet of semiconductor material can be influenced by applying a bias between the semiconductor and an electrode opposite to its surface.

ductance of the surface region changes by the field effect, the measurements are preferably done on thin samples of high resistivity so that the constant bulk conductance is small.

It appears that the field effect on the conductance is often less than calculated from the induced charge. One reason for this is that the mobility of the holes and electrons near the surface is decreased because the transverse electric field causes extra scattering of the charge carriers at the surface. Corrections for this effect have been calculated by Schrieffer [2-3]). But even when mobility corrections are taken into account, the surface conductance is often much lower than expected. This may then be explained by the presence of surface states in which the induced charge can be trapped.

It takes always some time before the induced charge is trapped, so that the primary effect of a change of the voltage between the metal and the semiconductor is always a change in surface conductance, which decays then afterwards. The time constant of the decay is a measure of the relaxation time of the surface states. It has been found that on etched germanium and silicon surfaces the time constants can range from less than 10^{-6} s to several hours. It is believed that the short relaxation times are due to centres near the interface between the crystal and the oxide layer present at the surface. These states are called "fast" states. The "slow" states (time constants more than 1 millisecond) may be due to centres in the oxide film (or on the outer side of it). Transport of electron or hole to these centres takes more time than transfer to interface states. It is also possible that certain slow-state effects are related to structural rearrangements at the surface due to the increased electron or hole concentrations and the electric field.

The states which are most active in changing charge with the valence or conduction band are those situated near the Fermi level. As the surface conductance is related to the position of this level between the valence and the conduction band, careful measurements can yield information both on the number and the energy distribution of the surface states.

When an inversion layer is present at the surface, the effect of an electric field on its conductance may be determined by measuring only between ohmic contacts applied to this layer. In fact the MOS transistor (figs 2.10 and 2.11), which was used in many instances in the work to be reported, is based on this principle. This type of measurement has the advantage that the conductance of the bulk material does not contribute to the measured conductance as this part of the sample is separated from the inversion layer by a depletion layer. On the other hand, the method has the severe limitation of giving only little information on surface conditions in which no inversion layer is present.

2.3.3. *The discovery of the stabilizing action of thermally grown oxide layers on the surface properties of silicon*

In 1959 a paper was published by Atalla, Scheibner and Tannenbaum [2-4]) dealing with the stabilization of silicon surfaces by thermally grown oxides. In this paper it was reported that the surface conductance, induced at an oxidized surface by a field effect did not depend on the ambient gas and remained constant over periods as long as 3000 hours. This indicated that the usual slow states observed on unoxidized surfaces had been eliminated, a conclusion of great importance.

Although ambient-gas and slow-state effects may thus be eliminated, this does not mean that there is no band bending at an oxidized surface and that no fast states are present. It was already indicated in Atalla's work that the surface properties depend on the method of oxide preparation and the presence of impurities. During the last years many publications have appeared in which it is concluded that thermal oxidation in either wet or dry oxygen always results in a more or less *n*-type surface (like that shown in fig. 2.2). Also the experiments to be described in this thesis show that in most cases donor centres at the surface predominate over acceptor centres.

2.4. The effect of the surface properties on semiconductor-device characteristics

Here we consider semiconductor devices, whose action is based on the presence of *pn* junctions, in particular diodes and transistors. The characteristics of the *pn* junctions, such as breakdown voltage and reverse leakage current, depend greatly on the doping levels of the *p* and *n* regions adjacent to the junction, whereas the presence of recombination centres for holes and electrons can also have some effect. At places where the *pn* junctions come to the surface, their properties may be influenced by the surface conditions.

The value of the breakdown voltage, for example, can be lowered due to surface effects. In fig. 2.4 this has been illustrated for a p^+n junction (p^+ means heavily doped *p*-type) in a sample in which the band bending is such that the

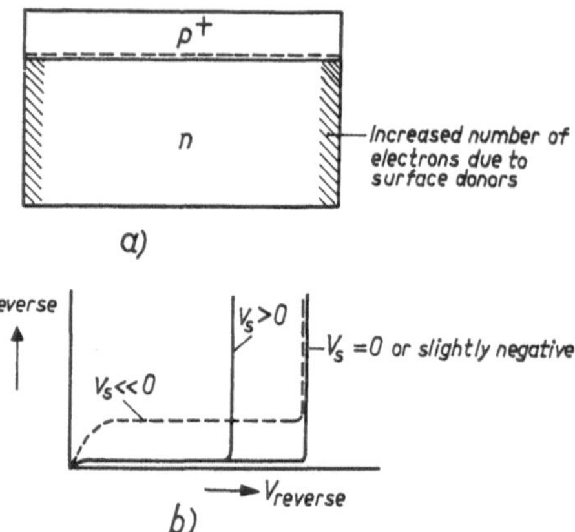

Fig. 2.4. Surface effects on the reverse I-V characteristic of a p^+n diode.
(*a*) Excess surface donors (i.e. the surface potential V_s is positive) cause the surface region of the n-type material to be of lower resistivity than the bulk. This situation occurs frequently in oxide-coated silicon diodes.
(*b*) When V_s is positive, breakdown effects in the reverse-biased junction occur first at the surface. The breakdown voltage is lower than expected from the doping levels of the p^+- and the n-type region. The breakdown voltage reaches a maximum when V_s approaches zero or is negative. However, for $V_s \ll 0$ the leakage current may be large due to formation of a p-type inversion layer at the surface of the n region.

n-type region is of lower resistivity near the surface than in the bulk *). As the p region is heavily doped, its properties are scarcely affected at the surface.

An excess of donor-type surface states (or positive oxide charge) on a sample with an n^+p junction may induce an n-type inversion layer at the surface (fig. 2.5). Although in such a case the breakdown voltage is not very greatly affected, electrons from the n^+ region may find a path through the inversion-layer channel towards regions (for example the edge of the wafer) where recombination with holes from the p-type side may occur readily. The current flowing under reverse bias conditions via a surface channel often shows a saturated character, which means that, although it may be large, it becomes independent of the voltage after this has reached a certain value. Such saturated current-voltage characteristics are also typical of MOS transistors, whose action is based on conduction through an inversion layer. The MOS transistor will be discussed in sec. 2.6.

Apart from an effect of the band bending, the presence of surface states may also cause an increased generation and recombination of holes and electrons at the surface and in this way influence the diode and transistor characteristics. The action of a junction transistor is based on the principle that charge carriers injected from the emitter into the base region can be collected at the reverse-biased collector-base pn junction. Any current which flows directly from the

*) In structures made by the "planar" technique (sec. 2.5) field crowding at the curvature in the pn junction (which is essentially not plane in this technique!) can be another serious reason for low breakdown.

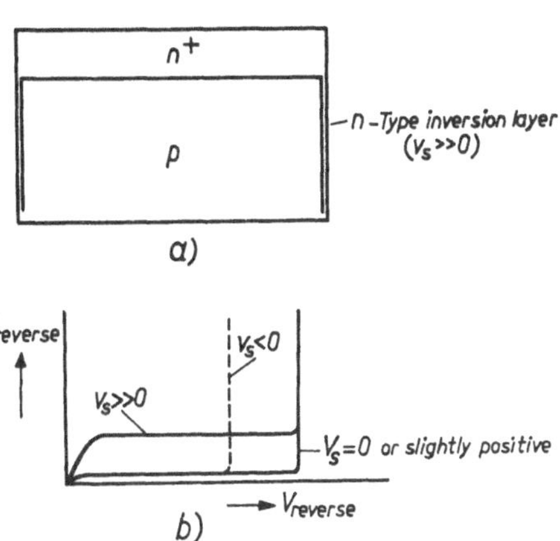

Fig. 2.5. Surface effects on the reverse *I-V* characteristic of an n^+p diode.
(*a*) Surface donors (often present in the case of oxidized silicon) may cause the presence of an *n*-type inversion layer on the *p*-type region.
(*b*) The presence of an inversion layer ($V_s \gg 0$) causes excess leakage of the *pn* junctions. This would be prevented by making V_s less positive. However, for $V_s < 0$ the breakdown voltage would be affected (compare fig. 2.4).

emitter to the base contact causes a decrease in amplification. Consequently channels on the base layer and surface recombination effects can affect the amplification properties.

Apart from the fact that it is often difficult for the diode or transistor maker to treat the surface in such a way that the properties of the device are as good as possible, another problem arises when the surface is to be put into such a condition that its properties do not change during the further life of the device (including tests at elevated temperature). In a large number of cases a reasonable stability can be obtained by a suitable etching or other surface treatments and a suitable choice of the substance in which the devices are encapsulated. In many cases the best stabilization method for a certain type of transistor or diode has been found by trial and error.

Surface stabilization is thus an important problem in semiconductor-device technology. Since transistors and diodes were first made, people have been looking into this problem. The finding of Atalla et al. [2-4]) that the surface properties of thermally oxidized silicon can indeed show a much better stability than those of the etched surfaces must therefore be considered as an important step forward. However, the fact that nowadays SiO_2 films are used very extensively in the manufacturing processes of silicon devices is not only due to their ability (when prepared well) to improve and stabilize the surface properties of the silicon, but also because they can be used as a mask against donor- or acceptor-impurity diffusion into the silicon. In the next section we will consider this property of oxide films and its consequences for the transistor technology.

2.5. The use of SiO_2 films for selective masking against impurity diffusion into silicon

2.5.1. *Diffusion of donor and acceptor impurities into silicon*

Diffusion of impurities into silicon is frequently used for making *pn* junctions, and this technique is also very suitable for making transistor structures with thin base layers. In the latter case two diffusion processes are applied, an acceptor diffusion followed by a donor-impurity diffusion for an *npn* structure (fig. 2.6)

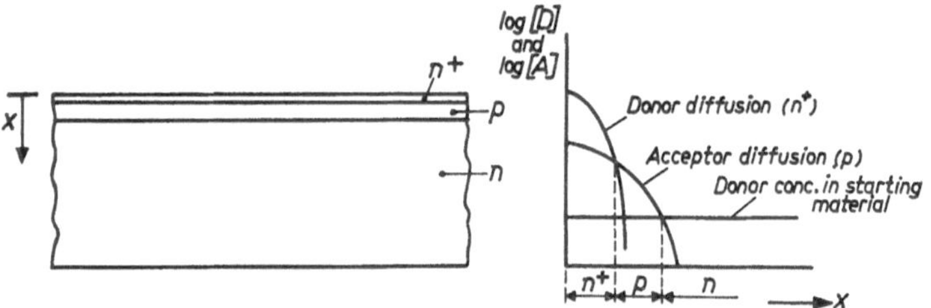

Fig. 2.6. Double diffusion processes can be used to provide transistor structures, in the illustrated case of the n^+pn type. After diffusion, it is difficult to make a contact to the thin base layer.

and vice versa for a *pnp* structure. Such diffusions can be carried out by heating the silicon sample in the vapour of the elementary impurity to be diffused or some compound of it. In practice phosphorus is most frequently used for donor diffusion and boron for acceptor diffusion. Both elements are generally offered in oxidized form, P_2O_5 and B_2O_3, respectively, and have to be reduced at the silicon surface. Consequently mixed-oxide layers ($SiO_2 + P_2O_5$ or $SiO_2 + B_2O_3$, often referred to in the literature as glassy layers) form at the surface. This results in protecting the silicon against evaporation, which might otherwise cause pitting of the surface. The surface concentration of the impurity in the silicon is related to the composition of these glassy surface layers. The composition of these and other surface layers formed on silicon during phosphorus diffusion from a P_2O_5 source will be considered in chapter 3.

2.5.2. *The masking action of* SiO_2 *films against impurity diffusion*

In the manufacture of transistors from diffused structures as shown in fig. 2.6, the difficulty of making a contact to the thin base layer is encountered. This contact can be made by local removal of the emitter layer and evaporation (and alloying) of a metal at the place where the base layer comes to the surface. Fortunately there is another, much handier, technique in which the emitter layer, and, when desired, also the base layer, can be diffused locally. This technique makes use of the masking action of SiO_2 films against diffusion, which was discussed in 1957 by Frosch and Derick [2-5]). Its principle is illustrated in fig. 2.7. A silicon wafer partly covered by an SiO_2 film is subjected

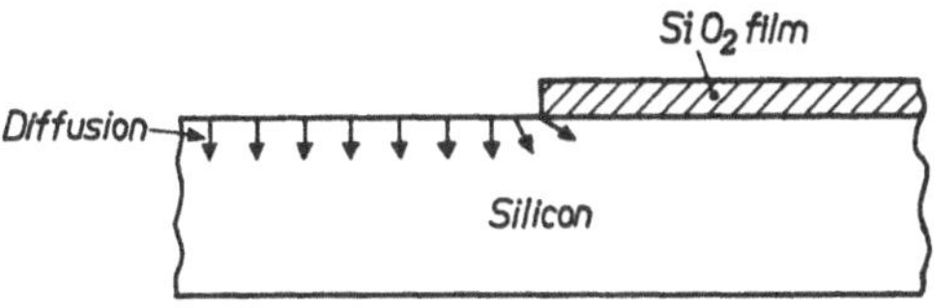

Fig. 2.7. The masking action of an SiO_2 film against impurity diffusion. By coating the surface locally with oxide, the diffusion may occur in selected regions.

to a diffusion treatment to introduce impurities into the silicon. The SiO_2 film may have been made by thermal oxidation of the silicon, after which it can be removed locally by applying an etch-resistant mask on parts of the film and subsequently etching in an aqueous HF solution. In cases like diffusion from a source of P_2O_5 or B_2O_3 the masking effect is accompanied by the formation of a new phase on the top of the SiO_2 film, consisting of a mixed oxide of SiO_2 with P_2O_5 or B_2O_3. During the diffusion period the thickness of this "glassy" phase may increase at the expense of the underlying SiO_2 ,and the masking action will stop as soon as the pure-SiO_2 film has disappeared. In chapter 3 the masking action will be considered in greater detail for the case of diffusion of phosphorus from a P_2O_5 source.

2.5.3. *The "planar" technology*

In the "planar" process [2–6]) diffusion and oxide-masking techniques are used to make *n* and *p* regions of limited area, but after the last diffusion step the oxide is not removed from the *pn* junctions, but only from a part of the diffused areas, to allow metal contacts to be made to the various regions. A schematic review of various steps necessary to make a "planar" *npn* transistor is shown in fig. 2.8. The structure of fig. 2.8*e* represents almost the complete stucture of a planar diode, only the metal contacts are not present. In many steps of the planar process use is made of photolithographic techniques, which make it possible to make several semiconductor devices from one slice of silicon. It is also possible to integrate various transistors and diodes in one piece of silicon, by connecting them by metal leads deposited on the oxide film. In such structures diffused regions may serve as electrical-resistance elements. In fig. 2.9 an example of a (rather simple) "solid circuit" is shown. This circuit, in which 3 transistors and 2 resistors are incorporated, is built in a piece of silicon with a surface area of $0{\cdot}6 \times 0{\cdot}6\ mm^2$. Several of such circuits may thus be made in a slice of silicon with a surface area of a few cm^2.

2.6. The MOS transistor

The discovery that silicon surfaces can be stabilized by thermally grown SiO_2 films has also made it possible to construct transistors which work on the

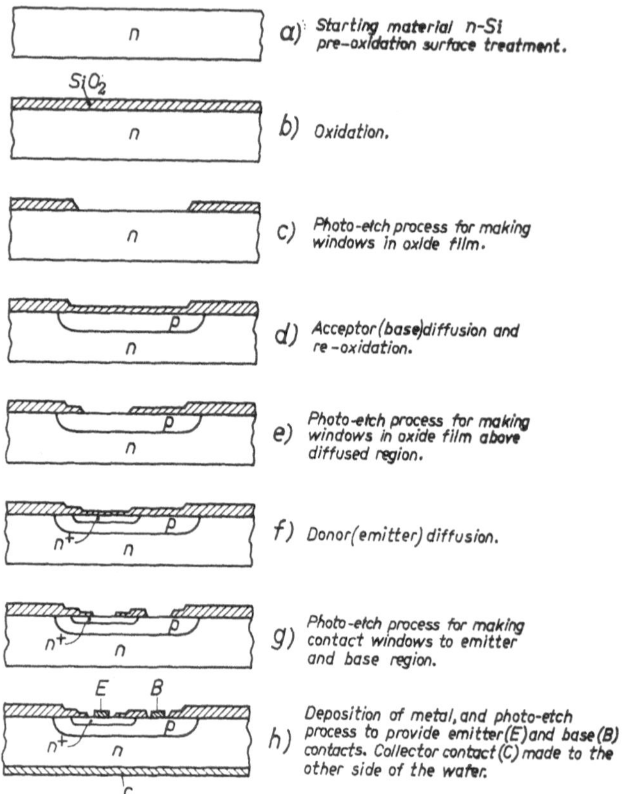

Fig. 2.8. Preparation of an *npn* planar transistor.

principle that the conduction of an inversion layer at a semiconductor crystal can be modified by a field effect. When a metallic field electrode is made on top of the oxide film, these types of devices are indicated as MOS (metal-oxide-semiconductor) transistors. In this section we will consider the principle of their operation and the influence of surface imperfections on their properties.

2.6.1. *Basic construction and operation*

The basic structure of the MOS transistor is shown in fig. 2.10. It consists in fact of two *pn*-junction diodes made in one piece of silicon, the surface of which is covered by an oxide film [2–7]). A metal electrode on top of the oxide covers the region between the *pn* junctions (there may be some overlapping across the junctions). A conducting path (i.e. an inversion layer) between the two *pn* junctions can be induced or modified by applying a bias to the metal electrode (the gate electrode) with respect to the semiconductor substrate. The conducting path or channel is separated from the substrate by a depletion layer. Two

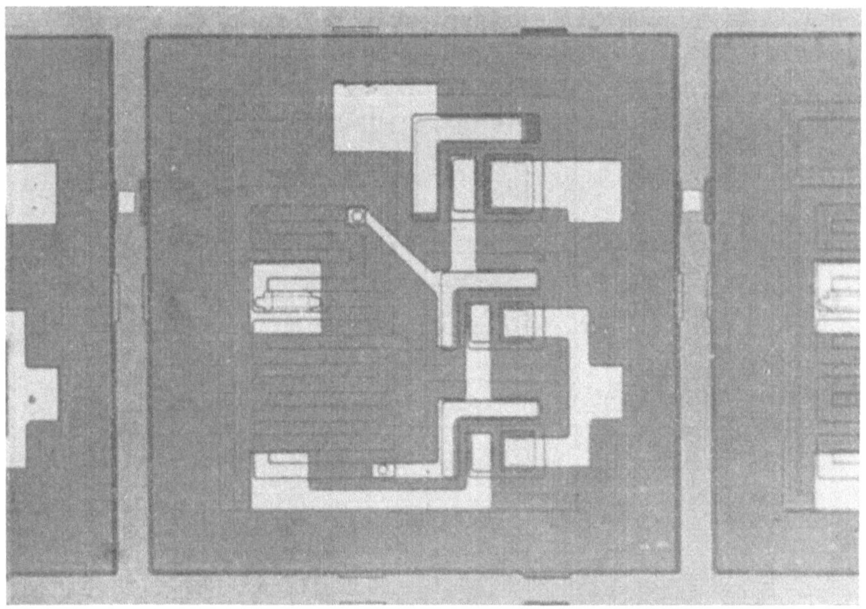

Fig. 2.9. An example of an integrated circuit in a silicon crystal; 3 transistors and 2 diffused resistors are interconnected by aluminium leads across the oxide film (this photograph was kindly made available by A. Schmitz of the Philips Research Labs).

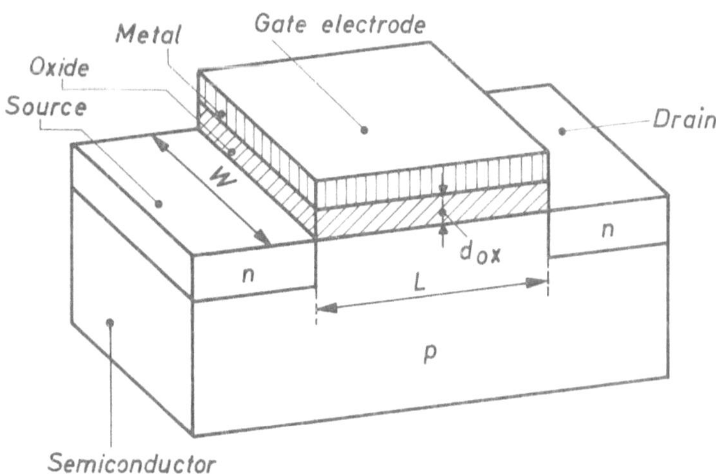

Fig. 2.10. The basic structure of the MOS transistor; d_{ox} = thickness of the oxide film, L the channel length and W the channel width (see also fig. 2.11).

constructions are possible, the *npn* (*n*-channel) MOST (fig. 2.11*a*), in which the channel conductance is increased when the gate electrode is made more positive and the *pnp* (*p*-channel) MOST (fig. 2.11*b*) in which a *p*-type inversion layer may be created by making the gate voltage negative.

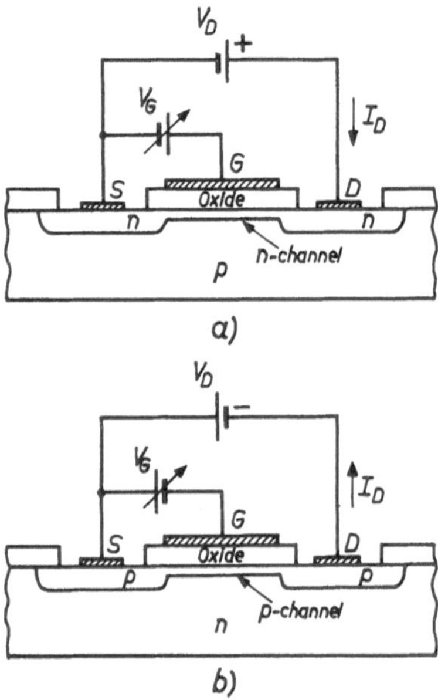

Fig. 2.11. (*a*) A cross-section through an *npn* (*n*-channel) MOS transistor. The drain current I_D can be varied by modifying the gate voltage V_G.
(*b*) A cross-section through a *pnp* (*p*-channel) MOS transistor. Note that the polarity of the drain voltage is different from that of the *n*-channel MOST.

The *pn* junctions are generally made by diffusion, using oxide-masking techniques. In general the diffused regions are heavily doped in contrast to the substrate, so that the field effect on the surface conductance is comparatively small in the diffused regions. The oxide film, used for masking against diffusion, may be used as insulating gate material, but after diffusion the oxide film may also be removed and replaced by another one.

In practice, a voltage is applied between the two diffused regions, so that one of them is reverse biased with respect to the substrate. This region is called the "drain", the other one the "source" (see figs 2.10 and 2.11). Leads are made to these regions as well as to the gate electrode. A separate electrode may be attached to the substrate, so that the device has four terminals. It is then possible to influence the channel conductance between source and drain by the voltage between gate and source, but, when an inversion layer is present, its conductance can also be lowered by reverse biasing the *pn* junction, i.e. by applying a reverse bias between source and substrate.

In many practical applications, the MOS transistor may be used as a three-terminal device. The bulk is then directly connected to the source. In the measurements to be mentioned in this thesis this was always done.

2.6.2. *The saturation of the drain current*

When there is a voltage difference (V_G) between gate and source so that an inversion layer forms between source and drain, this conducting channel will be homogeneous only so long as the voltage V_D between drain and source is zero. However, as soon as this is applied, the voltage between the gate electrode and the drain will be smaller than that between the gate and the source. Less charge will then be induced at the silicon surface near the drain. Since the drain region meanwhile becomes reverse biased with respect to the substrate, the inversion near the drain decreases for that reason too. When V_D is increased to a sufficiently high level there will not be any inversion at all near the drain: the channel is "pinched off". The value of V_D at which this just occurs is called the pinch-off voltage V_P. Due to the pinch-off effect the drain current I_D finds the highest resistance on its way immediately adjacent to the drain (fig. 2.12).

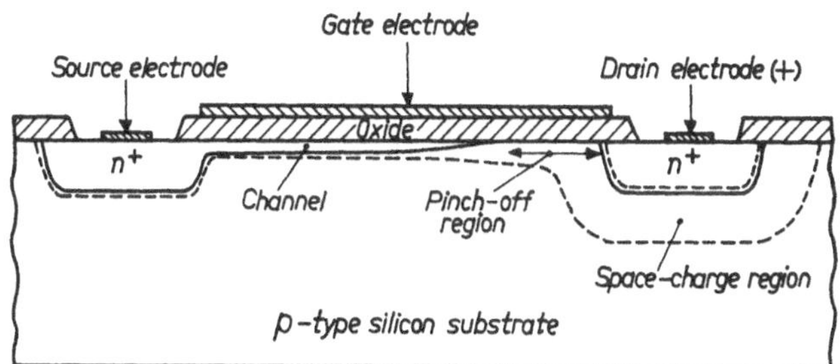

Fig. 2.12. The pinch-off effect in the channel of a MOS transistor occurs near the reverse-biased drain-substrate *np* junction. The space-charge region is very thin near the source, which is assumed to be connected to the substrate.

Consequently the largest part of V_D occurs across this region. When V_D is increased beyond V_P, all the extra voltage is taken up by this region. The current can still pass due to the high electric field parallel to the surface in the pinch-off region, and its value can be considered as being determined by the voltage drop and the resistance of that part of the channel which is not pinched off and is therefore nearly constant. This saturation effect of the drain current is shown in the I_D-V_D characteristics such as given in fig. 2.13.

2.6.3. *Relationships between the drain current and the gate voltage*

We will first consider the (theoretical) case that there is no influence of surface states and oxide charge. The drain current has then to be considered as a function of the type and dope of the semiconductor substrate, of the dimensions of

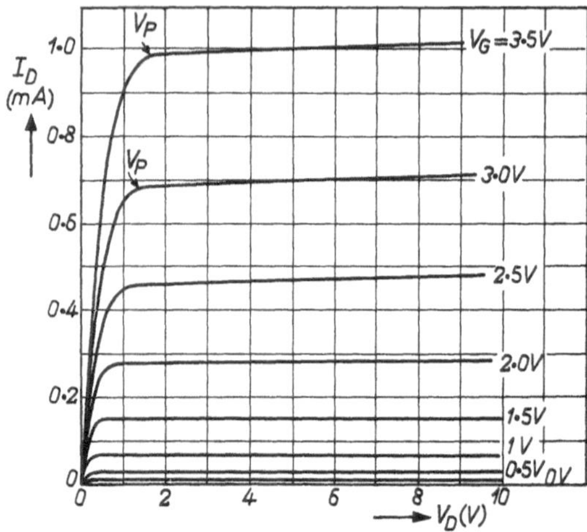

Fig. 2.13. An example of the drain characteristics of a MOS transistor; V_P is the pinch-off voltage, i.e. the drain voltage (V_D) where the drain current (I_D) starts to saturate at the given gate voltage (V_G).

the MOS transistor and the potentials of the gate, the drain and the source electrode (the latter is assumed to be connected to the substrate). The relationship between I_D and V_G will be considered first for small values of V_D, at which the pinch-off effect described in the previous section may be neglected, and then for values of V_D where this effect causes the drain current to be saturated.

2.6.3.1. The channel conductance at low values of the drain voltage

At sufficiently low values of V_D the current I_D is given by the channel conductance G_0, which is directly related to the number of charge carriers available for conduction in the inversion layer and further determined by their mobility and the dimensions of the MOS transistor:

$$\frac{I_D}{V_D} = G_0 = Ne\mu \frac{W}{L}, \tag{2.8}$$

where N is the number of contributing charge carriers per cm^2 surface area and μ their mobility; e is the unit charge ($1{\cdot}6.10^{-19}$ coulomb). The channel width W and length L are defined in fig. 2.10. N is only a part of the total number of charges present per cm^2 at the silicon surface (cf. fig. 2.2). The quantity of immobile charge due to depletion of majority carriers depends on the doping level of the substrate. Relations between the mobile charge in the inversion layer Q_{inv} and the total space charge Q_{Si} are given in fig. 2.14 for various doping levels (based on calculations by Whelan [2-8])). It can be seen that a certain charge

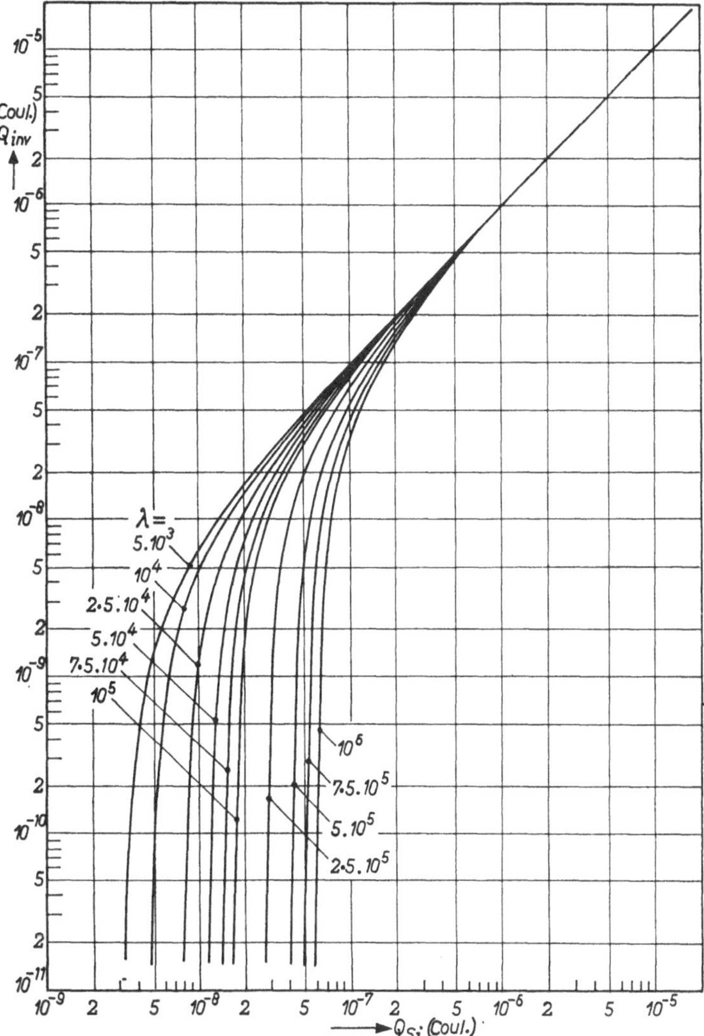

Fig. 2.14. The charge Q_{inv} in the inversion layer at the surface as a function of total space charge Q_{Si} for various doping levels of the silicon substrate ($\lambda = p_0/n_i$(p-Si) or n_0/n_i(n-Si) with $n_i = 1{\cdot}6.10^{10}$ cm^{-3}).

always has to be induced before inversion starts. In fig. 2.15 the charge density (per cm^2) at an intrinsic surface is given as a function of the doping level of the silicon substrate. A threshold value V_t of the gate voltage below which no channel conductance can occur may be calculated from this lower limit of induced charge. Figure 2.15 shows values of V_t (compared to the flat-band voltage V_f, which has been assumed to be zero until now) for a MOS transistor

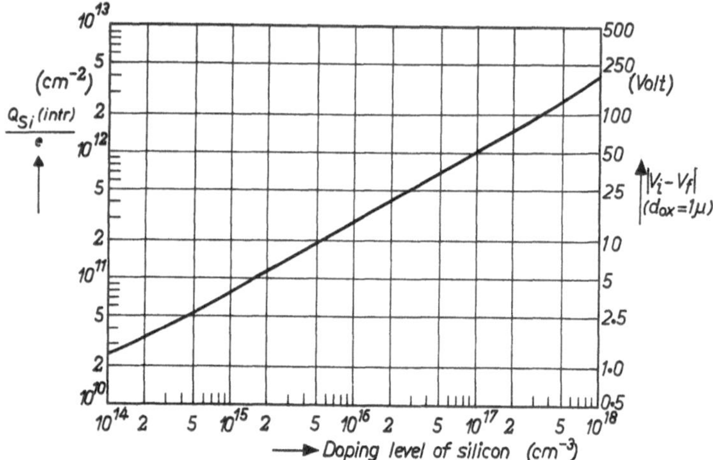

Fig. 2.15. The charge density (per cm^2) at an intrinsic silicon surface as a function of the doping level of the silicon substrate. In a MOS transistor at least this amount of charge has to be induced before channel conduction can start. The gate voltages V_i needed to get an intrinsic surface (compared to the flat-band voltage V_f) have been indicated for an oxide thickness of 1 micron.

with an oxide thickness of 1 micron. These have been calculated assuming that the charge density at the silicon surface is given by *)

$$Q_{\mathrm{Si}} = -C_{\mathrm{ox}} V_G. \tag{2.9}$$

When a certain inversion is established, a further increase of the gate voltage results mainly in an increase of the charge in the inversion layer, whereas the depletion charge can be shown to remain almost constant. The slope of a G_0-V_G curve is then independent of the doping level and can be determined by eq. (2.8), together with

$$e\frac{\mathrm{d}N}{\mathrm{d}V_G} = \frac{\mathrm{d}|Q_{\mathrm{inv}}|}{\mathrm{d}V_G} \approx \frac{\mathrm{d}|Q_{\mathrm{Si}}|}{\mathrm{d}V_G} \approx C_{\mathrm{ox}}, \tag{2.10}$$

giving

$$\frac{\mathrm{d}G_0}{\mathrm{d}V_G} = C_{\mathrm{ox}}\mu\frac{W}{L} = \beta, \tag{2.11}$$

with

$$\beta = C_{\mathrm{ox}}\mu\frac{W}{L} = \frac{\varepsilon_{\mathrm{ox}}\mu W}{d_{\mathrm{ox}}L}. \tag{2.12}$$

*) Equation (2.9) is only valid if the potential drop across the silicon space-charge region is negligible compared to V_G. The surface potential of silicon is generally not more than a few tenths of a volt. In fig. 2.15 it can be seen that for an oxide thickness of about 1 μ the threshold voltages calculated according to eq. (2.9) are mostly much larger indeed. However, for very thin oxide films and silicon with a low doping level the voltages across the oxide film and the silicon space-charge region may be of comparable value when the surface is just intrinsic.

For sufficiently large inversion, μ may be assumed to be constant (sec. 2.6.4) and then G_0 is a linear function of V_G.

2.6.3.2. MOS-transistor characteristics in the saturated region of the drain current

For values of the drain voltage V_D where the drain current saturates (I_D(sat)), the current is determined by the pinch-off conditions. For intrinsic material the pinch-off voltage V_P can be shown to be equal to the applied gate voltage V_G (with the aid of fig. 2.11 one can understand readily that the voltage across the oxide near the drain is zero when V_D equals V_G, which means that no inversion layer is induced at that place). The drain current can then be shown [2-9-11]) to be

$$I_D(\mathrm{sat}) = \tfrac{1}{2}\,\beta V_G{}^2\ (= \tfrac{1}{2}\,\beta V_P{}^2), \tag{2.13}$$

where the factor β is given by eq. (2.12).

The slope of the I_D(sat)-V_G line, often indicated as the transconductance (g_m) of the transistor, is thus dependent on the value of the drain current and is proportional to V_G (assuming again β, i.e. the mobility μ, to be constant):

$$g_m(\mathrm{sat}) = \frac{\mathrm{d}I_D(\mathrm{sat})}{\mathrm{d}V_G} = \beta V_G\ (=\beta V_P). \tag{2.14}$$

For non-intrinsic material the relationship between g_m(sat) and V_P remains valid, but V_P and V_G are no longer equal [2-11]). Instead, they may be related by the following equation:

$$V_P = F\,(V_G - V_0). \tag{2.15}$$

The parameters F and V_0 both depend on the doping level of the substrate. The factor F accounts for the fact that in doped material the pinch-off effect is caused partly by the reverse bias across the *pn* junction near the drain. For intrinsic substrate material F would be equal to unity, but it decreases when the material is more heavily doped with donors or acceptors. The value of V_0 relates to the part of the gate voltage which can be assumed to account for the immobile (depletion) part of the space charge at the semiconductor surface. As said before, the depletion charge becomes nearly independent of the gate voltage after a certain amount of inversion has been established.

The transconductance is now given by

$$g_m(\mathrm{sat}) = \frac{dI_D(\mathrm{sat})}{\mathrm{d}V_G} = \beta F\,(V_G - V_0) \tag{2.16}$$

and the drain current by

$$I_D(\mathrm{sat}) = \tfrac{1}{2}\,\beta F\,(V_G - V_0)^2. \tag{2.17}$$

For weak inversion V_0 is essentially a function of V_G, approaching the threshold value V_t when the surface is nearly intrinsic. At low values of the drain current the I_D(sat)-V_G curve cannot therefore be described by a parabolic law. The deviations become larger when the doping level of the substrate material is higher.

2.6.4. *The mobility of the charge carriers in surface channels*

A linear relationship between G_0 and V_G, resp. g_m(sat) and V_G (see eq. (2.11) and eq. (2.16)) is valid only when the factor β, i.e. the mobility μ, is a constant. In practice it appears generally that a large part of the G_0-V_G and I_D(sat)-V_G curves can indeed be described by a constant mobility.

In sec. 2.3 it was discussed that the effective mobility may be influenced by charge trapping in surface states and, at high transverse electric fields (strong inversion), also by a surface scattering mechanism. Surface states become generally less effective in decreasing the effective surface mobility due to charge trapping when the drain current is increased. This is due to the fact that in the case of strong inversion even a small change of the surface potential already causes a considerable change in the charge density in the inversion layer, whereas the occupation of the surface states is almost not changed. Hall-effect measurements on surface channels [2-12,13]) show that the mobility is indeed maximal for little inversion. Measurements of surface conductance cannot indicate this clearly, as the influence of surface states and depletion charge is then very large.

Recently, Arnold and Abowitz [2-14]) have reported electron mobilities close to the bulk values for surfaces of ⟨100⟩ surface orientation. These authors found a relationship between the mobility values (measured at strong inversion) and the number of interface states (electron-trapping centres). These centres are negatively charged when the surface is strongly inverted and may then decrease the electron mobility due to scattering effects. More generally one may probably state that the surface mobilities are highest for the most perfect crystal surfaces. Experimental results to be discussed in chapter 7 indicate that the Si-SiO_2 interface contains generally the smallest amount of defects for silicon surfaces of ⟨100⟩ orientation.

2.6.5. *The effect of interface states and oxide charge on the threshold value of the gate voltage*

It was stated in sec. 2.6.3 that a certain threshold gate voltage will be necessary to obtain a conducting channel in a MOS transistor. The value of the threshold voltage was shown to depend on the doping level of the substrate material (fig. 2.15). The threshold voltage V_i has been defined as the voltage necessary to obtain such a band bending that the surface is just intrinsic. In practice, however, it is more convenient to define a threshold voltage V_T as the gate voltage necessary to induce a certain drain current. This level may range from 10^{-10} to 10^{-4} A (in chapters 4 and 5, V_G at $I_D(\mathrm{sat}) = 10^{-5}$ A is often used as a reference point).

Firstly we may compare figs 2.16 and 2.17, which show the drain characteristics of *n*-channel MOS transistors made on the same material and under the

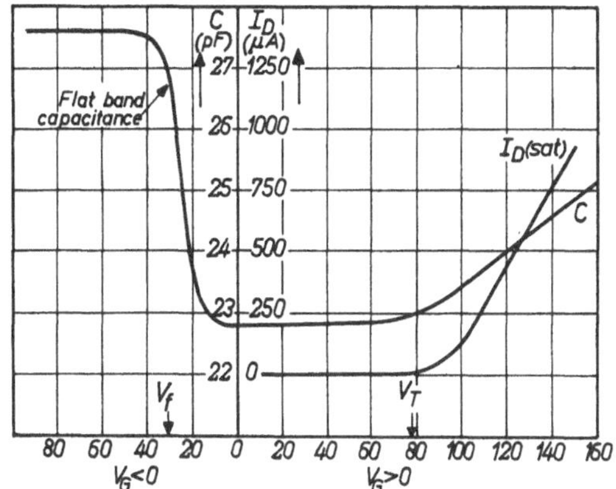

Fig. 2.16. Drain current (I_D) and MOS capacitance (C, see sec. 2.7) at 500 kc/s versus gate voltage (V_G) of an n-channel MOS transistor in which many interface states are present. This is indicated by the shape of the C-V_G curve and the large difference between the threshold voltage V_T of the drain current and the "flat-band" voltage V_f. The MOS-transistor structure is given in fig. 4.1.

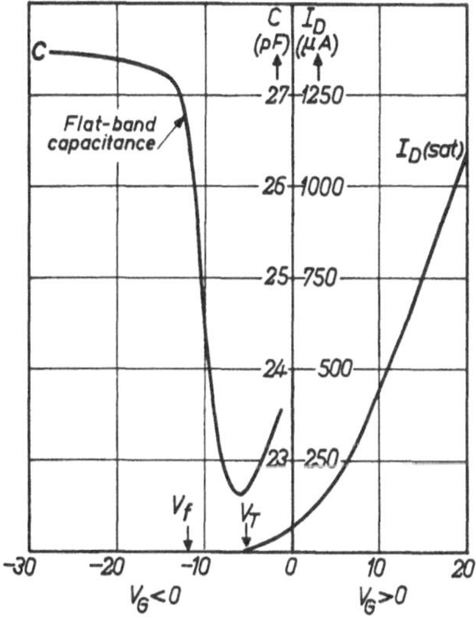

Fig. 2.17. I_D-V_G and C-V_G (at 500 kc/s; see sec. 2.7) curves of a similar MOS transistor as that of fig. 2.16. The number of interface states is made negligible as a consequence of a treatment of the oxidized slice in wet nitrogen at 450 °C.

same conditions, except for one additional heat treatment at 450 °C in wet nitrogen in the case of fig. 2.17. Apart from a difference in slope of the I_D(sat)-V_G curves, the threshold voltage appears to be quite different, about 80 V in fig. 2.16 and about —5 V in fig. 2.17. For the type of material (5 ohm-cm p-Si) which was used and the oxide thickness (1·2 μm) one would expect (from fig. 2.15) a threshold voltage of about 7 V.

To explain the observed differences, we must consider the influence of centres at the Si-SiO_2 interface (interface or surface states) and centres in the oxide film in which charge may be present. In the following considerations we will assume that the oxide charge Q_{ox} (per cm^2) is present very close to the Si-SiO_2 interface (when charge is distributed through the oxide film it is possible to account for the effect which the oxide charge induces at the silicon surface by representing it by a sheet of "effective" oxide charge at the Si-SiO_2 interface).

In fig. 2.18 Q_{ox} is shown together with other possibilities for charge in a MOS

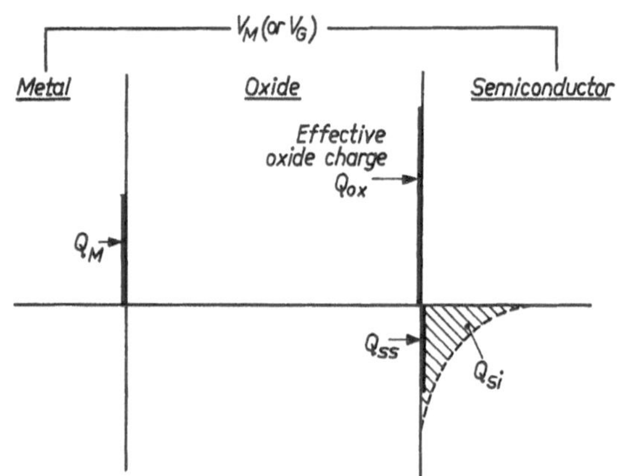

Fig. 2.18. An example of the charge distribution in a MOS system. The charge is distributed between the metal electrode (Q_M), the space-charge region in the silicon Q_{Si} ($Q_{Si} = Q_{depl} + Q_{inv}$; compare fig. 2.2), the surface states (Q_{ss}) and the oxide film (Q_{ox}, here assumed to occur only close to the Si-SiO_2 interface).

system: Q_M is the charge on the metal electrode, Q_{Si} that in the silicon space-charge region and Q_{ss} the charge in surface states (all per cm^2). Ignoring the voltage drop across the silicon space-charge region and the work-function difference between the silicon and the material of the gate electrode, one may say that a voltage V_M (in a MOS transistor V_G) between the metal (gate) electrode and the substrate induces an amount of charge on the metal electrode equal to

$$Q_M = C_{ox} V_M. \tag{2.18}$$

Both the charge on the metal electrode and the oxide charge can be assumed to induce charge at the Si-SiO_2 interface. This charge is distributed between the silicon space-charge region and the surface (interface) states:

$$Q_M + Q_{ox} = -(Q_{Si} + Q_{ss}) \tag{2.19}$$

or

$$C_{ox} V_M = -(Q_{Si} + Q_{ss} + Q_{ox}). \tag{2.20}$$

When the presence of surface states can be neglected, the presence of oxide charge causes a shift of the threshold voltage equal to

$$\Delta V_T = -\frac{Q_{ox}}{C_{ox}}. \tag{2.21}$$

As Q_{ox} is in most cases positive, this means a displacement of V_T in the negative direction *). Assuming for example that in the case of fig. 2.17 the presence of interface states can be ignored, the displacement of the threshold voltage from 7 V (theoretical case) to —5 V for an oxide thickness of 1·2 micron means the presence of an effective oxide charge equal to

$$Q_{ox} = -C_{ox}\Delta V_T = -\frac{\varepsilon_{ox}}{d_{ox}}\Delta V_T = -\frac{3{\cdot}4.10^{-13}}{1{\cdot}2.10^{-4}} \times -12 = 3{\cdot}4.10^{-8}\ \text{C}, \tag{2.22}$$

corresponding to about 2.10^{11} positive unit charges per cm^2.

There are strong indications (see for example chapter 7) that, in MOS structures in which the oxide is made by thermal oxidation, Q_{ox} always tends to be positive. It is reasonable to assume that this is the case for the transistor in both fig. 2.16 and fig. 2.17. In fig. 2.16, however, a high positive value of V_T is observed (in the order of 80 V), indicating the presence of immobile negative charge in centres at the surface. With Q_{ox} positive, this can only be explained by a negative value of Q_{ss} (see eq. (2.19); Q_{Si} is small at the onset of channel conduction). The transistor of fig. 2.16 should exhibit at least 10^{12} interface states per cm^2 in which electrons become trapped when the gate voltage is increased in the positive direction. Such surface states are shown in fig. 2.2 as acceptor-type interface states in the upper half of the energy gap. Donor-type interface states have been shown in the lower half of the gap. Presence of the latter may cause high negative values of the threshold voltage in *pnp* (*p*-channel)-type MOS transistors (see fig. 4.7).

The difference in slope of the I_D-V_G curves of figs 2.16 and 2.17 can also be explained by a difference in the number of interface states, but is probably also due to the fact that the charged centres affect the electron mobility (see previous section). The question why the transistor of fig. 2.17 shows much better characteristics than that of fig. 2.16 will be discussed in detail in chapter 4, briefly also in sec. 2.10.

2.6.6. *Effect of the electrode material on the threshold gate voltage*

When the voltage applied between the gate electrode and the silicon is brought

*) Excess donors in the silicon near the oxide may also cause a shift of V_T to more negative or less positive values. It is difficult to distinguish experimentally between these donors and donor states (positive charge) in the oxide structure; Q_{ox} thus incorporates the charge in both types of centres.

to zero, there is in general still a small potential difference given by the contact potential between the metal and the silicon. The potential difference depends on the type and resistivity of the silicon and the nature of the metal. The contact potential between *n*-type silicon of about 5 Ωcm and an aluminium gate contact may be assumed to be about zero. The use of aluminium in MOS transistors made on *p*-type material will then cause a shift of about 0·7 V of the I_D-V_G curve along the voltage axis in the negative direction.

The influence of the material of the gate electrode may be predicted from the work functions of the various metals which may be used for this purpose. In the literature there is considerable confusion about these work functions and also about the influence of various gate-electrode materials in MOS structures. When MOS structures are heated, reaction of the metal electrode with the oxide or with traces of water may result in the disappearance of surface states (chapter 4), whereas also ion-drift effects may ocur. Such effects depend on the nature of the metal and may be much greater than the influence of work-function differences.

2.6.7. *Some considerations on the design and manufacture of MOS transistors*

In the previous sections it was shown that the properties of MOS transistors depend on the dimensions, the substrate and gate material, and the heat treatments during or after oxidation which determine the amount of oxide charge and the number of surface states. The designer of MOS transistors has thus various possibilities to influence these properties.

A distinction can be drawn between *npn*- (*n*-channel) and *pnp*- (*p*-channel) MOS transistors, made on *p*-type and *n*-type material, respectively. Both types may be distinguished further into "enhancement" and "depletion" types. An enhancement-type MOST does not conduct at zero gate voltage. In a depletion-type device, however, there is already a conducting channel at $V_G = 0$. Taking an *n*-channel MOST as an example, the drain current can then be decreased by applying a negative gate voltage.

The presence of positive surface (oxide) charge is the reason why *p*-channel-silicon MOS transistors are in practice always enhancement mode (unless a *p* channel is made by an additional acceptor diffusion). In MOS transistors made on *p*-type material the positive oxide charge may be the reason why an *n*-type channel is already present at $V_G = 0$, so that these transistors are often of the depletion type. It is possible, however, to make the number of positive charges per cm^2 as low as 10^{11} per cm^2 or even lower (by suitable processing and preferably ⟨100⟩-surface-oriented silicon material (chapter 7)). As the doping level of the substrate material causes the presence of a certain threshold gate voltage compared to flat-band conditions (see fig. 2.15), enhancement-mode *n*-channel transistors may be made, provided the doping level of the *p*-type material is not too low. The enhancement-mode character may also be favoured when a metal with a high work function, e.g. nickel, gold or platinum, is used as the gate-electrode material instead of the more usual aluminium. However, the adhesion of the noble metals to the oxide is generally rather poor.

As was discussed in sec. 2.6.5, the positive oxide charge may also be compensated by negative charge in interface states. The threshold voltage can thus be affected by modifying the number of interface states (cf. for example figs 2.16 and 2.17, see also chapter 4). However, enhancement-mode n-channel devices prepared in this way may show instability due to slow-state effects[2-15]. Moreover, in order to obtain a high surface mobility, the sample should preferably be treated in such a way that the number of surface states is minimized.

The transconductance of a MOS transistor is determined by the factors $\beta(=\varepsilon_{ox}\mu W/d_{ox}L)$ and F in eq. (2.16). To obtain maximum transconductance n-channel ($\mu_n > \mu_p$) devices should be made on high-resistivity silicon (F approaches 1), with a thin oxide film. Further a small channel length and a large channel width should be used.

There are of course other requirements for a MOS transistor besides a large transconductance. In some instances the various requirements are conflicting and as there are also technological limitations in making components of small dimensions, the device made in practice is nearly always a compromise. In high-resistivity material, for example, the pinch-off region extends rapidly with the drain voltage and the remaining part of the channel thus rapidly decreases in length. Consequently the better transconductance for high-resistivity material is obtained at the cost of a decrease in the saturation resistance of the drain characteristic. The same occurs when the channel length is decreased. Compare for example the slopes of the curves of fig. 4.2 (channel length 400 microns) with those of fig. 4.5 (channel length 15 microns).

For applications in which the MOS transistor has to be used at high frequencies it is desirable to keep some of the capacities of the device low. This, for instance, means making the areas of the drain and gate region small and the oxide film thick. We have seen before that a thin oxide film is desirable for a high transconductance, and therefore the oxide above the channel region is still preferably made thin. Very thin oxide films also have the disadvantage that dielectric breakdown occurs even at a low gate voltage (the breakdown strength of a thermally grown SiO_2 film is about 5.10^6-10^7 V/cm) so that again a compromise has to be reached.

Practical MOS transistors have an oxide thickness between about 500 and 3000 Å, a channel width of a few hundred microns, and a channel length of only a few microns. In the work to be described in the chapters 4 to 6, use was sometimes made of MOS-transistor structures with a channel length of 400 μ and an oxide thickness of 1·2 μ. In these cases the MOS structures were prepared only to be used as a tool for measurement of surface properties of oxidized silicon.

2.7. Capacitance versus d.c.-voltage measurements on MOS structures

2.7.1. *The theoretical C-V curve (no surface states and oxide charge present)*

Measurements of the differential capacitance ($\mathrm{d}Q/\mathrm{d}V$) of MOS structures as a function of a d.c. voltage across it are often used, in this work too, to obtain information on the surface properties of oxidized silicon. The principle of the method will now be described qualitatively. Capacitance versus voltage measurements can be carried out on MOS-transistor structures, but also simply by applying a metal contact to an oxidized sample (fig. 2.19). A larger contact is made to the other side of the sample so that the contribution of the capacitance on this side to the total capacitance can be neglected. The high-frequency capacitance of such a MOS structure can be considered as the capacitance due

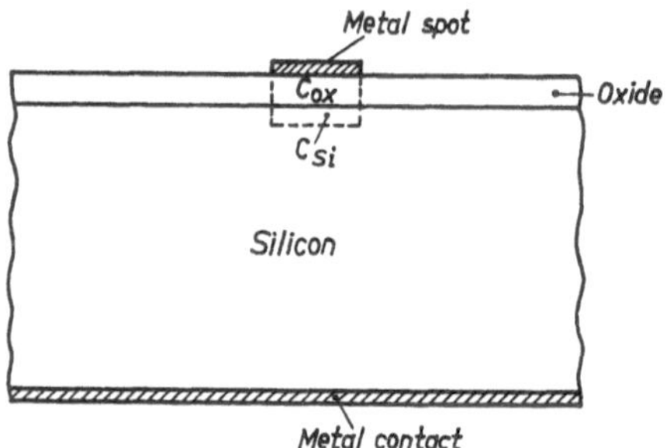

Fig. 2.19. Simple structure for MOS-capacitance measurements made by applying a metal contact to the oxide film. The other side of the wafer is provided with a larger metal contact, so that any contribution of this metal-semiconductor contact to the effective differential capacitance can be ignored.

to the oxide film C_{ox} in series with the space-charge capacitance C_{Si}, so that the total capacitance is given by

$$C = C_{ox}C_{Si}/(C_{ox} + C_{Si}) \tag{2.23}$$

(for ease of calculation C, C_{ox} and C_{Si} may be defined per cm^2 surface area). In this equation C_{ox} is a constant, but C_{Si} depends on the thickness of the space-charge region, which is related to Q_{Si}, the space charge per cm^2 and the impurity concentration in the material. As Q_{Si} depends on the voltage V across the MOS system and the oxide capacitance C_{ox}, theoretical C-V curves of ideal (no surface states and no oxide charge present) MOS structures can be calculated for each oxide thickness, if the impurity concentration in the silicon is known. The relationship between C_{Si} and Q_{Si} are ratheri ntricate, but have been published in graphical form by Whelan [2-8]), together with curves which relate these parameters to the surface potential V_s (indicated as $y_s = eV_s/kT$).

Typical shapes of theoretical C-V curves for MOS structures on p-type and n-type material are given in figs 2.20 and 2.21, respectively. Qualitatively, the shape of the curves can be explained as follows. Taking p-type material as substrate by way of example, a negative charge on the metal electrode will cause accumulation of holes at the surface. Variations in charge caused by the a.c. measuring signal then occur so close to the Si-SiO_2 interface that C_{Si} in eq. (2.23) is high compared to C_{ox}. The measured capacitance thus approaches C_{ox} for a negative voltage on the metal electrode. However, when the negative bias is low, i.e. when the band bending at the silicon surface is slight, the space-charge variations due to the measuring signal no longer occur solely close to the interface. Consequently C_{Si} becomes comparable to C_{ox}. This is particularly true for high-resistivity material and the consequence is that for zero and small negative voltages the effective capacitance can be considerably lower than the oxide capacitance. When the d.c. voltage at the metal electrode is increased from zero to positive values, a depletion layer forms at the silicon surface. An increase in

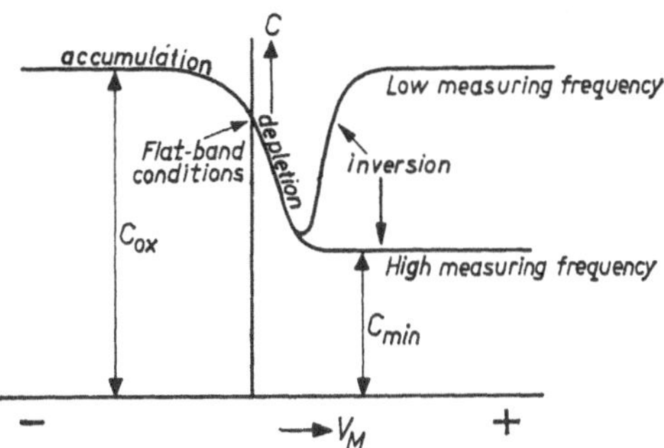

Fig. 2.20. *C-V* curve of a MOS structure on *p*-type silicon for high and low measuring frequencies; the oxide charge, the number of surface states and the work-function difference between the metal and the silicon are assumed to be zero.

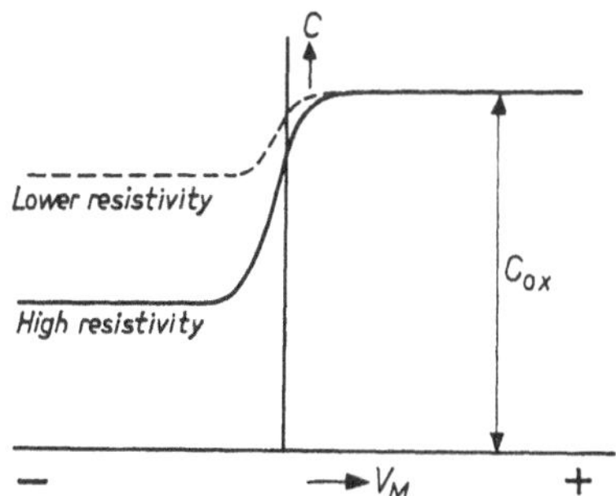

Fig. 2.21. High-frequency *C-V* curve of a MOS structure on *n*-type silicon under the same "theoretical" circumstances as in fig. 2.20. The effect of silicon resistivity on the minimum capacitance is also shown.

the negative charge in this layer is accompanied by an increase in thickness. The a.c. signal causes charge variations at the edge of the space-charge region. When the d.c. voltage is increased it looks as if the distance between the two electrodes of the MOS capacitor becomes larger, so that the capacitance decreases. However, at a sufficiently high voltage the band bending becomes so great that a layer of free electrons can be formed at the surface. As this inversion layer is very close to the Si-SiO_2 interface, the effective capacitance may again become equal to the oxide capacitance.

As sketched in fig. 2.20, a capacitance increase in the inversion region occurs only when the measuring frequency is sufficiently low (below about 100 c/s). At higher frequencies the supply of charge carriers to the inversion layer is difficult because these (minority) carriers cannot be generated quickly enough. In MOS-transistor structures, such as were employed frequently in the work to be described, the inversion layer is connected to the source and drain regions. An increase in capacitance in the inversion region may then also occur at high measuring frequencies. Supply of minority carriers may now occur via the diffused regions (moreover, the capacitance between these regions and the substrate is connected in parallel with the silicon space-charge capacitance when inversion occurs and the measured capacitance may then increase for that reason too).

An important part of the *C-V* curve is the accumulation-depletion region, where considerable change in the capacitance is observed. Each capacitance value between the maximum and minimum corresponds to a single value of the band bending. When there is no voltage difference between metal and silicon, the band bending is zero and the capacitance has some value between the maximum and the minimum. As sketched in fig. 2.21, both this "flat-band capacitance" and the minimum capacitance are closer to the value of the oxide capacitance when the material is of lower resistivity (for very low-resistivity material, i.e. when the semiconductor approaches the nature of a metal, hardly any change in capacitance occurs when the voltage is varied). The value of the minimum capacitance can therefore be used to determine the resistivity of the material. In practice the resistivity at the surface of oxidized silicon may differ from that in the interior of the crystal, so that an "average" resistivity of the material near the surface is obtained. Amongst the curves published by Whelan [2-8]) and which can be used to obtain rapid information from *C-V* measurements a useful graph is to be found from which the "flat-band capacitance" can be obtained relatively easy after the minimum and maximum capacitance have been measured. This graph, which was used frequently in this work to obtain the "flat-band voltage" V_f from a *C-V* curve, is reproduced in fig. 2.22.

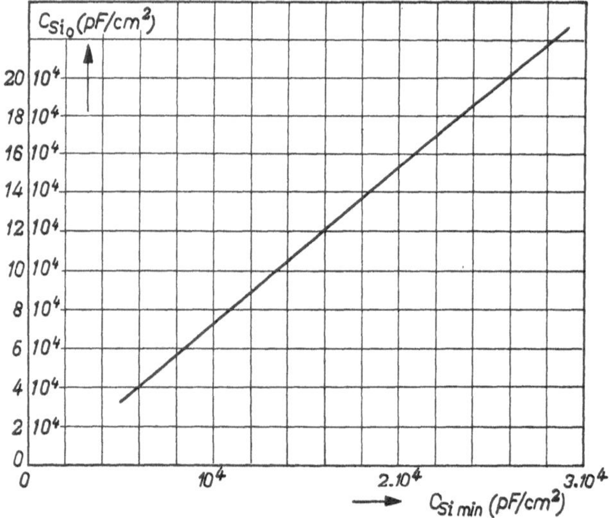

Fig. 2.22. The relationship between the minimum capacitance $C_{\mathrm{Si\,min}}$ and the flat-band capacitance C_{Si_0} of the silicon surface (according to Whelan [2-8])). This graph, in relation with the equation $C = C_{\mathrm{ox}}C_{\mathrm{Si}}/(C_{\mathrm{ox}} + C_{\mathrm{Si}})$, can be used to calculate the flat-band capacitance of a MOS structure after the maximum capacitance (C_{ox}) and the minimum capacitance have been measured.

2.7.2. *Displacement of a C-V curve due to surface charge*

A measured *C-V* curve often looks very similar to the theoretical one, with the exception that it is displaced along the voltage axis. For thermally oxidized silicon this shift is nearly always towards negative voltages due to the presence of a positive surface charge, either in the oxide or in the silicon near the surface. The shift is frequently expressed in a number of positive unit charges per cm^2 assumed to be present at the Si-SiO$_2$ interface (if the charge is distributed through the oxide layer, only an effective part is measured). In the *C-V* curve of the MOS transistor of which characteristics are given in fig. 2.17, the displacement (V_f) is equal to —12 V with an oxide capacitance of 27·2 pF for an area of the metal electrode of 1 mm^2. The effective number of positively charged surface centres per cm^2 is in this case thus equal to

$$N = -\frac{Q_M}{e} = -\frac{C_{ox}V_f}{e} = \frac{27{\cdot}2.10^{-10}.12}{1{\cdot}6.10^{-19}} = 2{\cdot}0.10^{11}\ \text{cm}^{-2} \qquad (2.24)$$

(the work-function difference $\Delta\phi_{MS}$ is ignored).

The same value was calculated earlier (sec. 2.6.5) from the displacement of the I_D(sat)-V_G curve.

2.7.3. *The C-V curve of a MOS structure in the presence of interface states and oxide charge*

When interface states are present, a change in the d.c. voltage across the MOS system may cause a change in their occupation. For donor-type surface centres this is determined by an equilibrium with electrons in the silicon:

$$D \rightleftarrows D^+ + e^-(\text{Si}) \qquad (2.25)$$

and similarly for acceptor-type centres by

$$A + e^-(\text{Si}) \rightleftarrows A^-. \qquad (2.26)$$

Various donor- and acceptor-type centres may be present, for each of which a certain energy level in the silicon band gap may be indicated. The total charge per cm^2 in these centres is represented in eq. (2.20) by the term Q_{ss}, which is equal to $e(N_{D^+} - N_{A^-})$ when N_{D^+} and N_{A^-} are the densities of ionized donor and acceptor centres per cm^2, respectively, and e is the positive unit charge.

Depending on the band bending and on the concentration and energy distribution of donor and acceptor states Q_{ss} may be either negative or positive. There will be a tendency for Q_{ss} to become negative or less positive at a positive value of V_M. Both the donor and acceptor states tend to be occupied by electrons; the donor centres are then not charged, whereas the acceptor centres tend to be negative. In a similar way it can be seen that Q_{ss} tends to become positive at high negative values of V_M.

As Q_{ss} is thus not a constant value but depends on the voltage across the

MOS structure, the experimental *C-V* curve may show a non-uniform displacement compared to the theoretical one. As each value of the capacitance corresponds to a certain band bending, the variation in the displacement can be used to obtain information on the distribution of the surface states in the silicon energy gap. However, the total amount of charge present in the states cannot be determined as a function of the band bending, as an unknown amount of charge may always be present in oxide centres. It is also impossible to determine by this method whether the states are of donor or acceptor type.

Detailed measurements of surface-state distributions with the aid of the *C-V* method have not been carried out by the author. However, such studies have been made by M. V. Whelan on surfaces which were oxidized and heat treated in the same manner as used for several experiments to be described in chapter 4. A set of *C-V* curves measured by Whelan [2-16]) on a phosphate-glass-covered oxide film on *n*-type silicon is reproduced in fig. 2.23. Especially at higher frequencies the slope is considerably different from that expected theoretically, pointing to presence of surface (interface) states. At lower frequencies the presence of interface states has obviously a much less pronounced effect. These results indicate that, in order to obtain information on presence of interface states,

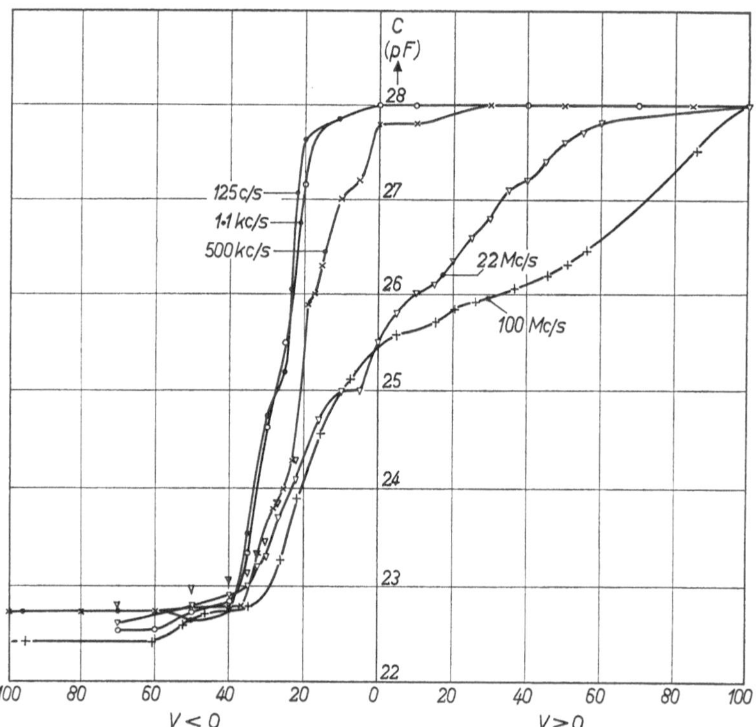

Fig. 2.23. *C-V* curves measured at various frequencies of a MOS structure on *n*-type silicon covered by an oxide film similar to that of the MOS transistor of fig. 2.16. Measured by Whelan [2-16]).

high measuring frequencies must be used. The reason is that it may be possible to establish the equilibria given in eqs (2.25) and (2.26) while a lower-frequency a.c. signal is applied, so that almost no change in the space charge Q_{Si} occurs. The capacitance value does then not correspond with the band bending at the given bias. Instead, a capacitance value close to the oxide capacitance may be measured.

In the work to be described in the following chapters the differential-capacitance measurements were always done at a frequency of about 500 kc/s. This frequency is too low to obtain much information about the presence of interface states from the slope of the accumulation-depletion part of the *C-V* curve. The voltage (V_f) at which the flat-band capacitance is measured does not necessarily correspond to real flat-band conditions either. One may argue, however, that *C-V* measurements in this lower frequency range give a useful indication of the amount of effective oxide charge (see sec. 7.4.1).

When a MOS transistor is considered, the *C-V* curve may be compared to the I_D-V_G curve. Both curves have been shown in figs 2.16 and 2.17 for two MOS transistors of the same dimensions, but with a difference in surface treatment. In the case of fig. 2.16 the difference between the threshold voltage and the "flat-band" voltage is large ($V_T \approx 80$ V, $V_f = -30$ V). When surface states are absent, such a large difference between V_T and V_f cannot be expected. In sec. 2.6.5 it was already argued that this MOS transistor exhibits more than 10^{12} interface states per cm^2. The *n*-type sample used for the *C-V* measurements indicated in fig. 2.23 had an oxidized surface which was treated in the same manner. These measurements have been used [2-16]) to get more information about the surface states and indicate that fast states (time constant in the order of 10^{-7} to 10^{-8} s) are lying about 0·1 to 0·3 V from the conduction-band edge. High-frequency *C-V* measurements on *p*-type samples with the same oxide show the presence of about the same number of states lying close to the valence band [2-16]). Calculations of capture cross-sections done by Whelan (private communication) indicate that the states near the conduction band are of the acceptor type (become negative after electron trapping), whereas the states near the conduction band are of the donor type (become positive after hole trapping). They have been indicated as interface states in fig. 2.2.

A comparison of the *C-V* curves of figs 2.16 and 2.17 shows further that, where many surface states are present (fig. 2.16), the capacitance curve shows a broad minimum. This is also an indication of the presence of surface states, as it means that the change from positive to negative band bending does not occur readily when the d.c. bias is varied.

It may be remarked finally that one must be cautious in interpreting every divergence of the slope of a measured *C-V* curve from the theoretical one as a surface-state effect. An inhomogeneous distribution of oxide charge over the surface can have a similar influence.

2.8. Preparation of silicon-dioxide films on silicon

2.8.1. *Thermal oxidation of silicon in oxygen and/or water vapour and its effect on the impurity distribution in the silicon substrate*

Thermal oxidation of the silicon substrate is the method which is most widely used for the preparation of oxide films for use as a mask against diffusion and passivation layer at the surface. Oxidation may be carried out either in oxygen, water vapour or a mixture of both. A simple oxidation apparatus is shown in fig. 2.24. This method is known as open-tube oxidation, a bubbler serving as

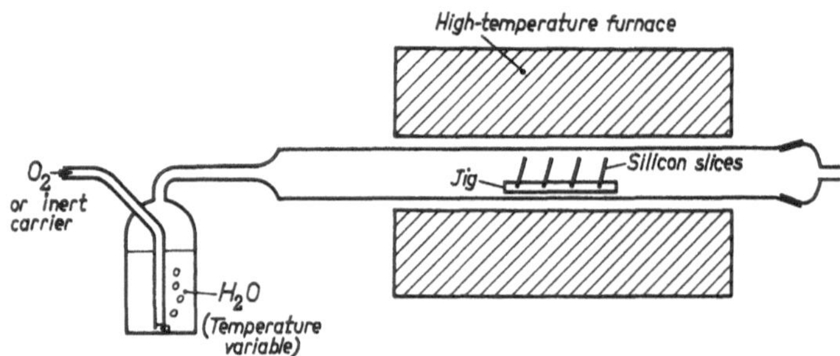

Fig. 2.24. Open-tube oxidation of silicon. When oxidation has to be carried out in dry oxygen, the water bubbler is omitted. The tube and the jig for holding the silicon slices are in general made of fused silica. Rate constants for various oxidation conditions are given in table 2-I.

the source of water vapour. The carrier gas may be either an inert gas or oxygen. The oxidation rate can be increased by increasing the water-vapour pressure, i.e. by increasing the temperature of the liquid water.

A general relationship for the oxide growth on silicon under these conditions has been given by Deal and Grove [2-17]). These authors conclude that the oxide thickness d_{ox} as a function of time t can be written as

$$d_{ox}^2 + Ad_{ox} = B(t + \tau). \tag{2.27}$$

The constants A and τ are only of importance for thin oxide films. For films of 2000 Å and more, grown at not excessively low temperatures (above about 1000 °C), the growth is limited by diffusion of the oxidant through the film and the growth can then be described by

$$d_{ox}^2 = Bt. \tag{2.28}$$

A number of rate constants for various oxidation conditions have been listed in table 2-I. In this table "dry" oxygen means that the gas contained less than about 5 ppm water. Oxidation in oxygen in a furnace system like that shown in fig. 2.24

may not be considered as oxidation in really dry oxygen because some water may diffuse from outside through the hot parts of the quartz tubes. Gregor and Balk [2-21]) report that by rigorous exclusion of water oxide growth is much slower than indicated by the growth rate for dry oxygen in table 2-I. This indicates some catalytic action of traces of water in the oxidation process.

As thermal oxidation is most effective in the same temperature range (above

TABLE 2-I

Rate constants for the thermal oxidation of silicon in an apparatus like that shown in fig. 2.24. The oxidation is assumed to occur following the general relationship [2-17]) $d_{ox}^2 + A\, d_{ox} = B(t + \tau)$ or, when no values for A and τ have been given, following the simple parabolic relationship $d_{ox}^2 = Bt$, where d_{ox} is expressed in microns ($1\ \mu = 10^{-6}$m)

ambient gas	temperature (°C)	$A(\mu)$	$(B \times 10^{-4})$ (μ^2/min)	τ(min)	reference
dry oxygen	800	0·370	0·2	540	2-17
	920	0·235	1·0	84	2-17
	1000	0·165	1·95	22	2-17
	1000	—	1·5	—	2-18
	1100	0·090	4·5	4·5	2-17
	1200	0·040	7·5	1·5	2-17
	1200	—	7·4-7·6	—	2-18, 19
wet oxygen (H_2O at 25 °C)	1200	—	12-13	—	2-18, 20
	1300	—	22	—	2-20
wet oxygen (H_2O at 95 °C)	920	0·50	33	0	2-17
	1000	0·226	48	0	2-17
	1100	0·1	84	0	2-17
	1200	0·05	120	0	2-17
steam (boiling H_2O)	1000	—	55	—	2-18
	1200	—	159	—	2-18
	1200	—	130	—	2-19
wet argon (H_2O at 28 °C)	1200	—	5·8	—	2-19
wet argon (H_2O at 85 °C)	1200	—	121	—	2-19

about 1000 °C) where diffusion of donor and acceptor elements in the silicon can be carried out, it must always be realized that some redistribution of these impurities may occur. For example, *pn* junctions may be displaced, and near the oxide-silicon interface redistribution effects may occur because certain impurities (the acceptor elements Al, B, Ga, In) tend to move into or through the oxide, whereas others (the donor elements P, As, Sb) are piled up near the interface, because they are rejected by the growing oxide film. These effects thus tend to make the surface less *p*-type or more *n*-type than the interior of the crystal and cannot always be distinguished from effects of positive surface charge in other centres. A thorough discussion of the redistribution effects, based on MOS-capacitance measurements, has been given by Grove et al. [2-22]).

During thermal oxidation, a considerable amount of oxygen may diffuse into the silicon. Although it is known that oxygen in silicon can cause formation of donor-type clusters at low temperatures, there are no clear indications that the oxygen diffusion is of great influence on the surface properties of oxidized silicon.

There are some methods for making oxide films at lower temperatures so that redistribution of impurities by diffusion in the silicon does not occur. A number of these methods, which all have certain disadvantages too, will be described briefly in the next sections.

2.8.2. *Oxidation of silicon in high-pressure steam*

It has been found that oxide films of good quality can be made by oxidation at relatively low temperatures if the water pressure is increased to several atmospheres. Results of this oxidation method have been described by Ligenza [2-23-25]) for the temperature range between 500 and 950 °C and for steam pressures from 25 to 500 atm. Within a certain range of pressures and temperatures uniform films could be made, the growth of which appeared to be linear in time and directly proportional to the steam pressure. At excessively high pressures, however, the oxide film may dissolve in the steam phase, so that no growth occurs. The time-independent growth rate indicates that the rate-determining step in this oxidation method is the oxidation of the silicon at the Si-SiO_2 interface. This explains also that in this oxidation method the oxidation rate depends strongly on the crystal orientation and also on the impurity content of the silicon [2-24]).

We have carried out a number of oxidation experiments by placing the silicon samples together with a calculated amount of water in a small metal bomb, similar to that described by Ligenza. Oxide films with good insulating properties were obtained, whereas the oxidized surface always showed an *n*-type character. The claim of Ligenza [2-25]) that this oxidation method can result in silicon surfaces with a *p*-type character when a certain pre-oxidation cleaning treatment was given, was not confirmed.

2.8.3. *Accelerated oxidation of silicon in the presence of* PbO

This method was reported by Kallander et al. [2-26]). The temperature necessary for oxide-film formation can be lowered by rate-accelerating additives. Lead oxide especially has appeared to be a suitable accelerating agent, and may be added through the vapour phase.

We have done a number of experiments using this method and found that the effects were greatest when a PbO source was placed close to the silicon sample to be oxidized. For a PbO-Si distance of 0·2 mm and an oxidation temperature of 650 °C the oxide thickness appeared to increase linearly with time with a rate of 0·01 micron per minute, for oxidation periods between 10 and 100 minutes. The rate can be influenced by the temperature, the oxygen pressure and the PbO-Si distance during oxidation. At 625 °C and the same PbO-Si distance the rate was about half that at 650 °C. When in experiments at 650 °C the PbO-Si distance was doubled, the oxidation rate decreased to about a quarter.

The great influence of the distance from the PbO source to the silicon may be explained by considering that the oxide layer which grows on the silicon does not consist of pure SiO_2 but must contain a certain amount of PbO, to the presence of which the oxidation rate is related. Considering the oxide film thus as a lead glass, it is apparent that the vapour pressure of PbO at the surface of the lead glass is lower than the PbO pressure near the PbO source. The amount of PbO transported to the silicon is thus determined by a diffusion process.

From a number of other experiments it became apparent that a certain lower limit of PbO has to be present to guarantee accelerated oxidation at a given temperature. When samples which had been subjected to accelerated oxidation were further heated in oxygen without any PbO source, some further oxidation appeared to occur. After some time (a few minutes at 650 °C), however, no further noticeable oxide growth occurred. When the temperature was increased, again some increase in thickness could be observed. Apparently the PbO content necessary for accelerated oxidation decreases with increasing temperature. Analysis of the film with an X-ray electron-probe microanalyzer showed that at 650 °C the limiting PbO content was in the order of 20 mol. %. The results point to a relatively easy diffusion of oxygen through the lead glass, whereas the oxidation rate is limited by the oxidation process at the silicon surface or by the rate with which the SiO_2 formed is "dissolved" in the lead glass. Below a certain PbO content (determined by the temperature) the latter probably no longer occurs. A very thin SiO_2 layer between the lead glass and the silicon will be sufficient to stop the oxidation of the silicon completely at the relatively low temperature.

The properties of silicon surfaces covered by these lead-containing oxide films appeared to depend greatly on heat treatment. Although in many cases almost intrinsic surfaces were obtained, it appeared to be possible to make the surface either strongly *n*-type or *p*-type by certain heat treatments (sec. 2.10.1.2).

Lead-glass coatings may also be prepared by evaporating PbO or Pb on Si with or without an SiO_2 film on its surface followed by heating in an oxidizing environment.

2.8.4. *Growth of* SiO_2 *films from a reaction between silicon halides and water*

In this case and also in the methods to be discussed in the next sections the oxide film is not made by oxidizing the silicon substrate, but is deposited on the substrate. It is well known that SiO_2 can be made from a reaction of $SiCl_4$ and H_2O:

$$SiCl_4 + 2\,H_2O \rightleftarrows SiO_2 + 4\,HCl. \tag{2.29}$$

Steinmaier and Bloem [2–27]) have shown that this reaction is very suitable for the preparation of SiO_2 films on silicon, especially when it is combined with epitaxial growth of silicon from the vapour phase. In such an epitaxial process a silicon layer is deposited on a single-crystalline substrate (generally silicon) by heating this substrate in a mixture of $SiCl_4$ and H_2, so that a reduction of the $SiCl_4$ occurs at the hot substrate:

$$SiCl_4 + 2\,H_2 \rightleftarrows Si + 4\,HCl. \tag{2.30}$$

After a layer of silicon is deposited in this way, water vapour may be added so that SiO_2 forms, Instead of H_2O it is convenient to inject CO_2, which results in formation of water according to the reaction

$$CO_2 + H_2 \rightleftarrows H_2O + CO. \tag{2.31}$$

In the experiments of Steinmaier and Bloem the growth rate of SiO_2 increased from about 0·2 μ per hour at 1000 °C to about 100 μ per hour at 1350 °C.

According to Rand and Ashworth [2–28]) SiO_2 films can be prepared at a lower temperature (800-850 °C) if $SiBr_4$ is used instead of $SiCl_4$. The method may then also be used for making SiO_2 films on other semiconductors such as germanium.

The properties of oxide films made by the $SiCl_4$ method have appeared to be comparable to those of layers grown by thermal oxidation of silicon in dry oxygen. The surface charge is rather low (in the order of 2.10^{11} positive charges per cm^2).

It is possible to prepare doped oxide films by addition of suitable compounds (e.g. PCl_3, BCl_3) to the vapour phase during growth.

2.8.5. *Pyrolysis of organic silicon compounds*

A rather convenient method for preparing SiO_2 films at relatively low temperatures is heating the substrate (silicon) in the vapour of organo-oxy-silanes such as tetra-ethoxy-silane ($Si(OC_2H_5)_4$). Preparation of such films has been described amongst others by Klerer [2–29,30]) and Jordan [2–31]). Each oxy-silane has its own decomposition range, in general lying between 600 and 800 °C. When oxygen is added, SiO_2 films can be grown at temperatures as low as 350 °C.

The films do not consist of pure SiO_2, but still contain a number of organic groups. Their structure is less dense than that of oxide films made by thermal oxidation. Densification may be carried out after the deposition by heating at more elevated temperatures [2–32,33]).

2.8.6. *Deposition of* SiO_2 *films by vapour-transport methods*

It is possible to make SiO_2 films by vapour deposition. However, very high temperatures are needed to evaporate pure SiO_2, so that sputtering processes are felt to be more suitable. Evaporation of the more volatile silicon monoxide (the solid SiO must probably be considered as a mixture of Si and SiO_2) in systems containing oxygen or water may lead to the formation of films with a composition somewhere between SiO and SiO_2, depending on the growth conditions [2-34]).

Chu and Gruber [2-35]) have described a chemical transport method of depositing SiO_2 films or silicon, using hydrogen fluoride as the transport agent in a closed system. The silicon substrate was heated to 400-600 °C, the SiO_2 source to a lower temperature.

Until now the deposited types of SiO_2 film mentioned in this section have not found major applications as coatings directly on the silicon. Films made by thermal oxidation appear to yield better surface properties.

2.8.7. *Anodic oxidation of silicon*

Silicon can be oxidized anodically in a suitable electrolyte, although the ionic-current efficiency of film growth is low. For *n*-type silicon the anodic current is in the reverse direction for a surface barrier on the silicon, and the rate of oxide growth is limited by the supply of minority carriers, holes, to the Si-SiO_2 interface [2-36]). Illumination with light causes an increased oxidation rate and local illumination allows of local oxide growth. Schmidt et al. [2-37,38]) have shown that certain donor and acceptor impurities may be incorporated in the SiO_2 film by a suitable choice of the electrolyte. The doped oxide film may then be used as a source of impurity diffusion into silicon during subsequent heating.

Anodic SiO_2 films have not yet found many applications in semiconductor-device technology. Atalla [2-4]) states that films of this type have a much larger content of ionic impurities than films made by thermal oxidation, which may make them less suitable as a passivating agent. However, this may depend on the nature of the electrolyte which is used. Schmidt [2-39]) has pointed to a difference in the stress present in oxide films made by thermal oxidation or by the wet anodic process. In the former case the oxidation reaction proceeds via oxygen diffusion and the films are under strong compressive stress, partly because of the difference between the coefficients of thermal expansion of SiO_2 and Si. However, anodic films are formed by a cation-migration process [2-37]) and show no stress at all or even tensile stress. It might well be that the stresses in oxide films have an influence on the structure of the Si-SiO_2 interface, but at the moment there are no indications for it.

Silicon can also be oxidized by making it the anode in a gas discharge, a method which was described by Ligenza [2-40]). Considerable oxidation rates were obtained at temperatures as low as 300 °C.

2.9. Chemical and physical properties of thermally grown SiO_2 films

The phase diagram of SiO_2 is shown in fig. 2.25 as given by Singer [2-41]).

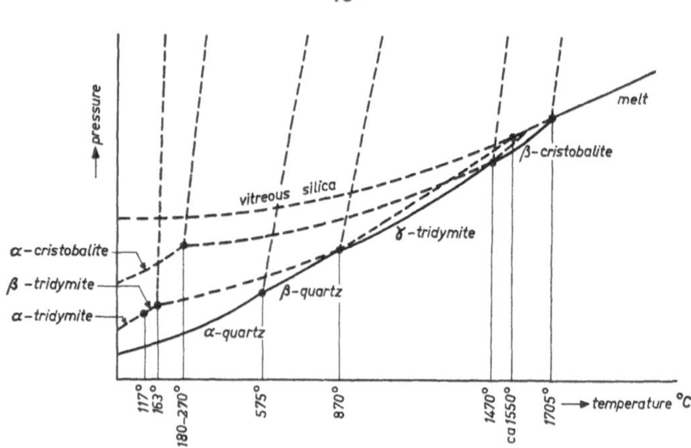

Fig. 2.25. The phase diagram of SiO_2, according to Singer [2–41]).

Several crystal modifications can exist, each consisting of a regular network of SiO_4 tetrahedra. Fused silica may be considered as an undercooled liquid with a more or less random network of SiO_4 groups. During heating of fused silica crystallization may occur, particularly in the range 1000-1200 °C if certain impurities are present.

The oxide films which are made by thermal oxidation of silicon are often considered to have a fused-silica-like structure, i.e. a random network of SiO_4 groups. When oxidation has been carried out in dry oxygen, a considerable degree of order may exist in the Si-O network. Crystals may be observed if the surface was not cleaned in a proper way before oxidation, an effect which is reminiscent of the devitrification of fused silica mentioned above. Some of these crystallites have been identified by electron diffraction as α-crystobalite [2–42]).

Silicon-oxide films containing many crystallites often show a comparatively low electrical resistance, probably due to leakage along the crystal boundaries. Clean and dust-free working is therefore essential for preparing oxide films of good quality.

An idea of the electrical conductivity of the films can be obtained by depositing metal spots on the oxide and measuring the current between metal and silicon when a voltage is applied. For good oxide films the resistivity is higher than 10^{16} Ωcm. When the voltage across the oxide is increased, dielectric breakdown may occur, generally when the electric field is in the order of 5.10^{6}-10^{7} V/cm, but at a lower field when weak spots are present. The method is thus suitable for obtaining an idea of thc quality of the film. From thickness measurements combined with capacitance measurements on MOS structures, the dielectric constant of the oxide film may be determined. For thermally grown oxides the dielectric constant appears to be about 3·8, a value which corresponds with that of fused silica or quartz. The refractive index of the oxide films has been found [2–43]) to be about 1·46. These figures and also infrared-transmission spectra correspond quite well with those of fused silica [2–33,42]).

The presence of pores in the oxide film can be detected by a chlorine etch technique [2-4]). The oxidized samples are heated in chlorine gas, e.g. at 900 °C, for a couple of minutes. The chlorine does not attack the SiO_2, but penetrates through the pore and etches the silicon, thus undercutting the SiO_2 layer. The etched area can be seen later under a microscope. Etching in Cl_2 or HBr may also be used to remove the silicon substrate from the oxide layers. Ing et al. [2-32]) did gas-permeation studies on such SiO_2 films, from which they concluded to the presence of micropores of atomic dimensions.

Deal [2-18]) has observed that, in oxides formed by oxidation in steam, small pits and etched areas may occur. Moreover, in these cases the amount of SiO_2 present is not quantitatively equal to the amount which one would expect from the amount of silicon which disappears during the oxidation. This points to incomplete oxidation. In pure oxygen the "oxidation efficiency" was found to be 100%, in wet oxygen (H_2O at 95 °C) 95-100%, in steam (1 atm.) 75-100%. This means that in the latter cases the oxide layer has a lower oxygen content than corresponds with the formula SiO_2, or that some silicon is transported into the vapour phase. Claussen and Flower [2-19]) report differences of oxide growth obtained during oxidation in wet oxygen and wet argon. In the latter case the growth rate (at 1200 °C) appeared to be very low during the first minutes of oxidation. During oxidation in wet but oxygen-free environments some SiO may possibly form, because of the reducing action of the hydrogen due to reduction of water:

$$Si + 2\,H_2O \rightleftarrows SiO_2 + 2\,H_2. \tag{2.32}$$

In the early state of oxidation the concentration of hydrogen will be fairly high near the Si-SiO_2 interface and the formation of SiO is favoured:

$$Si + H_2O \rightleftarrows SiO + H_2. \tag{2.33}$$

These phenomena may also be described due to reduction of SiO_2 by H_2:

$$SiO_2 + H_2 \rightleftarrows SiO + H_2O. \tag{2.34}$$

When oxygen is also present, SiO will not form so easily and, as soon as an oxide film is present, the reaction will slow down, so that the hydrogen concentration decreases and formation of volatile SiO is no longer a serious effect.

It is interesting to note that Edagawa et al. [2-42]) have found by infrared-absorption measurements that oxide films made in steam contain a region near the silicon surface which does not contain as many SiO bonds as may be expected from an SiO_2 composition. This may again be considered as being due to the reducing action of hydrogen formed during oxidation.

The hydrogen content of oxide films is of course much larger for oxidation in wet gas than for oxidation in oxygen. The hydrogen is probably mainly present in the form of OH groups, although possibly partly also in SiH groups or as dissolved molecular hydrogen or water. Hydrogen concentrations as high as 10^{20} atoms per cm^2 in steam-grown oxides have been measured [2-45]).

Differences in properties between various oxide films may also be detected from etch-rate studies. A suitable etch solution was developed by Pliskin [2-33]). This "P-etch" consists of 15 parts hydrofluoric acid (49%), 10 parts nitric acid (70%) and 300 parts water. As examples of the sensitivity of this etch to packing density, the etch rate of pyrolytic films formed at 675 °C is 13 Å/s, whereas the etch rate of thermally grown SiO_2 films is only 2 Å/s at 25 °C. Pliskin and Lehman [2-33]) report etch rates of 20-70 Å/s for electron-gun-evaporated silicon-oxide films and 4 Å/s for oxide films made by a reaction of $SiCl_4$ and H_2O at high temperatures. Mixed oxides of P_2O_5 and SiO_2 dissolve rapidly (300-500 Å/s) in the P-etch, which can therefore be used to determine the thickness of the phosphate glass present at the top of an SiO_2 film after this is subjected to a heat treatment in P_2O_5 vapour.

To answer the question whether an oxide film prepared by a certain method will be suitable for surface stabilization of silicon devices it is not sufficient to know the density, the refractive index and the etch rate of the layer. One would prefer a density which is as high as possible to guarantee a good isolation of the silicon surface from the ambient atmosphere. However, ion transport may also occur in more dense SiO_2 films under the influence of an electric field, as a consequence of which oxide-covered devices may show instabilities. It is therefore important also to have information about the presence of small concentrations of impurities in dependence on the way of oxide preparation. A number of measurements and considerations about this subject can be found in chapter 7 (see also sec. 2.11).

2.10. Effect of oxidation and further treatments on the properties of the Si-SiO_2 system

In the fabrication of "planar" devices, any oxidation and diffusion process does not necessarily provide a good device. Furthermore, when a certain oxide coating has been proved to be suitable for a certain type of diode or transistor, this does not necessarily mean that it can be used for another type of device too. It is felt that in most cases the surface properties are determined by the charged centres in the oxide and interface states at the oxide-silicon interface. How these properties can be influenced by various processes is discussed extensively in the chapters 4 to 7. The main conclusions of this work will now be reviewed.

2.10.1. *The effect of the oxidation process and further heat treatments on the surface properties*

2.10.1.1. Interface states

After thermal oxidation of silicon a considerable number of surface states can be present. These states are probably due to unsaturated silicon bonds at the Si-SiO_2 interface. They are able to capture holes or electrons depending on their concentration at the silicon surface. Consequently in MOS transistors, both of the *n*-channel and the *p*-channel type, a high gate voltage may be necessary to start conduction through the inversion layer (see for example fig. 2.16), whereas the transconductance can be very low.

Fortunately there are methods of influencing the concentration of these interface states. The surface orientation of the crystal appears to be of great influence. The lowest numbers, for a given oxidation procedure, are found when the surface consists (more or less completely) of a ⟨100⟩ plane. More states are found on a ⟨110⟩ plane and again more on a ⟨111⟩ plane [2-46]).

It appears that the number of interface states depends in a rather intricate manner on the water content of the ambient gas during oxidation and further heat treatment. It is proposed that the hydrogen component of the water plays an important role in this behaviour. At elevated temperatures hydrogen may create unsatured silicon sites at the Si-SiO_2 interface due to its reducing action on the oxide structure (impurities like sodium may cause a similar effect, see sec. 2.10.1.2 and chapter 7). On the other hand, a reaction with the unsatured bonds may result in the formation of SiH bonds, so that surface states disappear. SiH bonds are not very stable at high temperatures. Consequently treatments of oxidized silicon in water vapour or hydrogen are most favourable for the disappearance of interface states when the temperature is below about 600 °C. When the oxide is in an oxygen-deficient state (heat treated in an inert gas at elevated temperature) only traces of water are needed to cause the effect. Low-temperature treatments in wet nitrogen or hydrogen may be used in order to obtain a good transconductance in MOS transistors (compare for example fig. 2.16 with fig. 2.17). It can be dangerous to remove all interface states in oxide-covered n^+p diodes or p^+np-junction transistors, as residual oxide charge may then induce an inversion layer on the *p*-type material.

The annihilation of surface states during a low-temperature heat treatment can also occur in an inert gas if a reactive metal contact (for example aluminium) is present on the oxide film. The experiments indicate that this is probably due to a reaction of the metal with hydroxyl or water in or at the top layer of the oxide film, so that hydrogen atoms can diffuse towards the Si-SiO_2 interface. This effect can cause difficulties (channels) during heat treatment (for example the contacting procedure) of devices in which metal-oxide-silicon structures occur. This is the case, for example, in silicon solid circuits, where vapour-deposited aluminium is often used as a material for leads across the oxide film.

When an oxidized sample is subjected to a P_2O_5 diffusion in a dry environment, the number of interface states can increase considerably. This is probably due to the fact that oxide becomes very dry under these circumstances. It appears that the presence of a phosphate glass on top of the oxide film makes the introduction of water (or hydrogen atoms) into the oxide film during low-temperature treatments much slower. As such the phosphate glass has a stabilizing action on the properties of the $Si\text{-}SiO_2$ interface (the phosphate glass decreases also the possibility for certain ions to drift into the oxide film; this effect will be considered in sec. 2.11.4).

Although the work described in this thesis refers mainly to diode and MOS-transistor characteristics, it may be noted that there are clear indications that the amplification factor of planar junction transistorsis also dependent on the number of the indicated surface states [2-47]. Treatments which are known to improve the quality of MOS transistors may also be used to increase the current-amplification factor of oxide-coated silicon transistors, especially at low current densities, where surface-recombination effects can have considerable influence. The same treatments may also decrease the noise in planar transistors. For MOS transistors a relationship between $1/f$ noise and surface states has recently been indicated [2-48].

2.10.1.2. Oxide charge

The surface properties of oxidized silicon cannot be described by the presence of interface states alone. Several experiments point to a likelihood of positive charges being present in the oxide film, although it is in general difficult to distinguish positive oxide charge experimentally from donor centres with a low activation energy at the $Si\text{-}SiO_2$ interface. The positive surface charge may be related to the presence of unsaturated silicon bonds in the oxide structure near the silicon surface. The number of these bonds depends on the presence of impurities such as hydrogen and sodium. In particular when sodium is present, the surface-charge density is much higher for $\langle 111 \rangle$ than for $\langle 100 \rangle$ silicon surfaces. Oxidizing treatments, especially when carried out at relatively low temperature, generally cause an increase of the charge and treatments in an inert environment a reduction. Low-temperature treatments which can decrease the number of interface states (see previous section) can also cause a decrease of the positive surface charge. These and other experimental results may be explained by a model of the $Si\text{-}SiO_2$ interface structure, to be discussed in chapter 7.

The amount of surface charge in oxide films made by thermal oxidation of silicon in the presence of PbO (sec. 2.8.4) can be very low. It has been observed [2-44] that in this case the surface charge can even be made negative when a temperature gradient is applied across the oxide-silicon system in such a way that the oxide is "cool" with respect to the silicon. This effect is most readily explained by accepting that the mixed lead-silicon oxide contains centres which are able to trap electrons. Under the described circumstances electrons may tend to flow to the oxide. Similar effects, although much smaller, were observed with oxide films grown in steam. In these cases, however, the oxide charge always remained positive. The effects are a warning against making too quantitative conclusions from measurements on oxidized silicon. During the cooling process after thermal oxidation or further heating, temperature gradients may occur readily.

2.10.2. *Effects of ionizing irradiations*

Various effects of ionizing irradiation are described in chapters 5 and 6 in which it is shown that a tendency for electron transport from the silicon to the oxide can be created with irradiation by u.v. light. The photon energy of the light should be at least 4·2 eV. However, whether the electrons can be trapped there so that the positive oxide charge is decreased, depends on the structure of the oxide film. Such a decrease appears to be hardly possible after oxidation in dry oxygen, but considerable effects can be observed when the oxide can be supposed to contain water or hydroxyl groups. It is difficult to say whether the trapping centres were present already before or induced during irradiation, as it may be expected that hydroxyl groups can be attacked by u.v. light.

X-ray irradiation tends to cause an effect opposite to that of u.v. light, i.e. the positive oxide charge increases. This effect can be explained by the creation of free electrons in the SiO_2 which are "trapped" in the silicon, i.e. they need an energy of several electronvolts to go back from the silicon surface to the oxide. This reverse transport is helped when the X-ray irradiation is followed by illumination with u.v. light. Then it may turn out that the X-rays have created trapping centres in the oxide film: the u.v. effect is often increased compared to before X-ray irradiation. Creation of centres in the oxide film (and at the Si-SiO_2 interface) is most effective when the oxide is grown in a wet environment.

The effect of ionizing irradiation can be influenced when an electric field is applied in a metal-oxide-silicon system. The amount of oxide charge after such a treatment appears to depend on the properties of both the metal-oxide and the oxide-silicon interface (chapter 5).

Although ionizing irradiation can thus have a great influence on the surface properties of oxidized silicon, these effects are not of particularly great practical importancc for controlling the surface properties, as the induced effects disappear during heating above 200 °C. Because, however, the irradiation effects depend greatly on the method of oxide preparation, they can be used as a tool to detect differences in the oxide-silicon system after various treatments.

2.11. Stability problems in oxide-coated silicon devices

In the previous section it was indicated that changes in the surface properties of oxidized silicon may occur during ionizing irradiations. Oxide-covered silicon devices, particularly those containing MOS structures with a voltage across it, may therefore show considerable instability under these conditions.

Instability effects may also be observed in the absence of ionizing irradiation. Not every oxide film is suitable for the protection of every type of device to be made. The properties of an oxide-protected silicon surface depend on the energy distribution of surface states at the Si-SiO_2 interface, on the field due to the charge present in or on top of the oxide film and on the donor and acceptor distribution in the silicon near the interface. Any change in these parameters during the life of the device may cause instability of its properties. It can be assumed that the impurity distribution in the silicon does not change during a life test, which is generally carried out at or below 300 °C, although some donor formation due to the presence of oxygen may occur under certain circumstances. Various reasons for instability will now be considered in more detail.

2.11.1. *Slow-surface-state effects*

As was already indicated by Atalla [2–4]), such effects are generally not observed in oxidized silicon surfaces. They may be noted after ionizing irradiation, indicating that this can induce oxide traps in which electrons or holes from the silicon surface may be trapped with relatively long time constants. Usually, slow trapping effects are not observed in MOS transistors, because slow states are either not present or cannot be observed due to ion-drift effects which work in the opposite direction and may be greater. The absence of slow-state effects point to a fairly perfect, in any case trap-free, oxide structure near the silicon surface. It may be interesting to remark that we have observed slow-state effects occasionally when an imperfect interface structure was also indicated by the presence of a large number of fast surface states. These observations may correspond with certain slow-state effects mentioned in the literature [2–15]). We have found, however, that by suitable treatments (chapter 4) both the fast- and the slow-state effects can be minimized. It is thought that this is due to the reaction of unsaturated silicon bonds at the interface and in the oxide with hydrogen.

2.11.2. *Charge motion on the top of the oxide film*

In 1959 Atalla et al. [2–49]) described instability phenomena in oxide-protected silicon *pn* junctions under reverse bias. The leakage current of *pn* junctions can increase markedly under these conditions, especially when the ambient gas contains moisture. The phenomena were thought to be due to displacement of ions on top of the oxide due to the fringing field of the reverse-biased *pn* junction. Shockley et al. [2–50]) did experiments under similar conditions and observed clearly that a positive charge forms above the *p* region and a negative charge above the *n* region.

One may explain this in the first instance by considering the oxide surface as a "conductor" (i.e. the conduction at the top of the oxide, whatever it may be due to, is large compared to conduction in the oxide). When a reverse bias is applied across an underlying *pn* junction, the "conductor" on top of the oxide tends to be charged in a manner as indicated in fig. 2.26: it forms a capacitor with respect to both the *p*- and *n*-type region. Because oxide-covered silicon

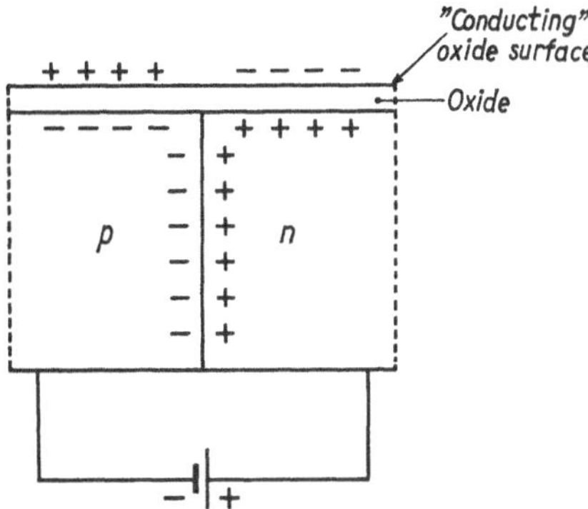

Fig. 2.26. A reverse bias applied across an oxide-coated *pn* junction tends to cause charge redistribution atop the oxide. When the oxide surface is "conductant", e.g. due to the presence of moisture, negative charge forms above the *n* region and positive charge above the *p* region. Charge of opposite sign is then induced at the silicon surface. This may result in the formation of inversion layers at the *p*- and/or *n*-type region with the consequence of large leakage currents.

surfaces have already a tendency to be *n*-type, the surface of the *p* region may readily be inverted. The surface of the *n*-type region tends to become less *n*-type and, when the charge becomes high enough, even *p*-type. Consequently the effects may cause an increase in the leakage current of an n^+p junction (*n*-type channel on the *p* region), and an increase in the breakdown voltage of a p^+n junction, eventually followed by an increase in leakage current (*p*-type channel on the *n* region, or *n*-type channel on the p^+ region).

The effects indicate that it is desirable to have oxide-coated silicon devices encapsulated as dry as possible. Where this is not possible or wanted, a thick oxide film should be used. This may also be achieved by depositing SiO_2 or another glassy film on the thermally grown oxide film. It may be remarked that the effect is also slight when the oxide is made in such a way that a large number of charge-neutralizing surface states is present at the Si-SiO_2 interface. The effect may be prevented by applying metal layers on the top of the oxide above the *p* and *n* regions and connecting them to these regions. The top layer of the oxide is then always at nearly the same potential as the underlying semiconductor.

Charge motion at the outer surface of the oxide film may also be induced by applying a voltage to a metal electrode partly covering the film. The oxide surface around the electrode then tends to obtain the same potential. In MOS transistors in which the gate electrode does not completely cover the region between source and drain, this effect can cause serious instabilities. When, for example, a conducting channel is induced by applying a bias across the MOS system, the conduction tends to increase in the course of time because charge leakage from the metal electrode over the oxide surface also tends to induce a channel in the region which is not covered by the metal electrode. The effect can be minimized by ensuring that there is little surface contamination and that the ambient gas is dry. In MOS transistors it can be prevented by ensuring that the metal electrode covers at least the whole area between source and drain region.

2.11.3. *Charge motion in the oxide film*

When mobile charge carriers are present in the oxide film, similar phenomena may occur during reverse biasing of a *pn* junction as described in the previous section for charge migration on top of the film. As ions move more easily at higher temperatures, these effects are increased when the temperature is raised. It has been shown that such effects can be caused by sodium contamination [2–51]).

In MOS structures the presence of a transverse electric field in the oxide film may result in shifts of the I_D-V_G and C-V curves along the voltage axis. To obtain a good idea of the influence of the charge-transport processes under these conditions, it is necessary to know which charge carriers in the oxide film can contribute to this transport and which electrode reactions may occur at the M-O and the O-S boundary. Presumably the number of free electrons in the SiO_2 film is very small (except when ionizing irradiations are applied) and charge transport in the oxide film may be assumed to be due primarily to ion migration. At the boundaries these ions may be neutralized or generated, but when neither of these processes occurs readily, space charge will be built up near the electrode where the charge transfer is most difficult. In silicon oxide univalent ions such as alkali ions can be assumed to be capable of migrating under the influence of an electric field. It has been found that contamination of the outer oxide surface with sodium ions can cause serious instabilities in MOS structures where the metal electrode is made positive with respect to the silicon [2–52]). The drift of sodium ions into the oxide can be speeded up by increasing the temperature. The increased positive oxide charge causes the C-V curve of the MOS structure to be displaced towards more negative voltages until an equilibrium situation is established in which the displacement of the C-V curve corresponds with the quantity of sodium deposited on the oxide surface. This indicates that the sodium ions are not easily neutralized at the Si-SiO_2 interface so that they accumulate there and form a layer of positive surface charge.

Contamination of thermally oxidized silicon with sodium may also occur during the various stages in the preparation of MOS structures. Contamination of the vapour-deposited metal electrode can be a serious cause of ion-drift effects. However, oxide films may also be contaminated by sodium in the furnace during oxide growth or further treatments (see chapter 7). The sodium concentration is then highest in the top layer of the oxide film. These sodium ions may also cause instability effects, especially when the metal electrode is made positive with respect to the silicon in a bias-temperature treatment of a MOS structure.

However, the effect of such a treatment is generally much lower than expected from the number of sodium ions being present, indicating that are they not all displaced very readily under the influence of an electric field.

Hofstein [2-53]) has pointed to the fact that the activation energy which may be measured from ion-drift effects as a function of the temperature is not a measure for the activation energy of the diffusion of the ions under the influence of the applied electric field. Instead, some dissociation or ion-generation effects at the metal-oxide interface determines the velocity of the drift phenomena. This is apparent from the fact that removal of the bias applied during an ion-drift experiment causes a very quick decrease in the induced oxide charge, pointing to a very fast ion diffusion in reverse direction. Moreover, the ion-drift effect in the forward direction can be increased considerably by the presence of traces of water.

It may be that in the presence of water protons may also cause ion-drift phenomena in MOS structures [2-53]), but a definite answer to this question is difficult to give at the moment. Only slight ion-drift effects may be observed when a MOS structure is heated with the metal electrode negative, even when the oxide film has been grown in a wet gas. This indicates that the hydrogen present in the oxide is tightly bonded, probably mainly in hydroxyl groups. Near the oxide-silicon interface often some SiH groups may be assumed to occur, and the hydrogen may then be considered to be present as "negative ions" (chapter 7). One might speculate that "neutralization" of this hydrogen can be the reason of the increase of the number of positive charges and interface states which may be observed to occur [2-54-56]) after heating MOS structures at 300 °C or higher with a negative bias applied to the metal, although explanations in terms of creation of positive silicon ions or oxygen vacancies may not be excluded.

2.11.4. *The stabilizing action of phosphate-glass layers*

Although contamination of MOS structures may be largely prevented by careful processing (chapter 7) it has been found that also contaminated oxide films may be stabilized by subjecting them to a P_2O_5 diffusion of some kind. A phosphate glass is then formed at the top of the oxide film. This phosphate-"glass" layer appears to getter sodium from the underlying SiO_2 and prevents the ion-drift effects observed in "bare" oxide films [2-57,58]). Another instability effect is introduced, however, because polarization effects may occur in the phosphate-glass layer [2-58]). These effects are not clearly understood but can be minimized by making the glass layer thin compared to the underlying SiO_2 and they depend further on the method of preparation of the phosphate glass. A phosphate glass with a low P_2O_5 content may be preferred in this respect. Methods of influencing the P_2O_5 content are discussed in chapter 3.

In conclusion it may be said that, by using careful processes, oxidized silicon samples and also MOS structures can be made which contain a minor number of oxide charges and interface states. From the point of view of most practical applications the surface properties can then be considered stable, except when the samples are subjected to ionizing irradiations.

REFERENCES

2-1) N. B. Hannay, Semiconductors, Reinhold Publishing Corp., New York, 1959.
2-2) A number of papers dealing with investigations on etched surfaces has been summarized in: R. H. Kingston, Semiconductor surface physics, University of Pennsylvania Press, 1956.
2-3) J. R. Schrieffer, Phys. Rev. **97**, 641-646, 1955.
2-4) M. M. Atalla, E. Tannenbaum and E. J. Scheibner, Bell Sys. techn. J. **38**, 749-783, 1959.
2-5) C. J. Frosch and L. Derick, J. electrochem. Soc. **104**, 547-553, 1957.
2-6) See for example: G. E. Moore in E. Keonjian (ed.), Microelectronics, McGraw-Hill, 1963, Ch. 5.
2-7) D. Kahng, U.S. Patent 3102230 (1960).
2-8) M. V. Whelan, Philips Res. Repts **20**, 620-632, 1965.
2-9) S. R. Hofstein and F. R. Heiman, Proc. IEEE **51**, 1190-1202, 1963.
2-10) C. T. Sah, IEEE Trans. **ED-11**, 324-345, 1964.
2-11) J. A. van Nielen and O. W. Memelink, Philips Res. Repts **22**, 55-71, 1967.
2-12) N. Murphy, Surface Science **2**, 86-92, 1964.
2-13) A. B. Fowler, F. Fang and F. Hochberg, IBM J. Res. Dev. **8**, 427-429, 1964.
2-14) E. Arnold and G. Abowitz, Appl. Phys. Letters **9**, 344-346, 1966.
2-15) M. M. Mitchell and N. H. Ditrick, Solid State Design **11**, 19-22, 1965.
2-16) M. V. Whelan, Philips Res. Repts **20**, 562-577, 1965.
2-17) B. E. Deal and A. S. Grove, J. appl. Phys. **36**, 3770-3778, 1965.
2-18) B. E. Deal, J. electrochem. Soc. **110**, 527-533, 1963.
2-19) B. H. Claussen and M. Flower, J. electrochem. Soc. **110**, 983-987, 1963.
2-20) C. T. Sah, H. Sello and D. A. Tremere, J. Phys. Chem. Solids **11**, 188-298, 1959.
2-21) P. Balk, Fall Meeting Electrochemical Society, 1965, Abstract 111.
2-22) A. S. Grove, O. Leistiko and C. T. Sah, J. appl. Phys. **35**, 2695-2701, 1964.
2-23) J. R. Ligenza, J. electrochem. Soc. **109**, 73-76, 1962.
2-24) J. R. Ligenza, J. phys. Chem. **65**, 2011-2014, 1961.
2-25) J. R. Ligenza, U.S. Patent 2930722 (1960).
2-26) D. A. Kallander, S. S. Flaschen, R. J. Gnaedinger and C. M. Lutfy, Spring Meeting Electrochemical Society, Extended abstracts, **10**, 129-130, 1961.
2-27) W. Steinmaier and J. Bloem, J. electrochem. Soc. **111**, 206-209, 1964.
2-28) H. J. Rand and J. L. Ashworth, J. electrochem. Soc. **113**, 48-50, 1966.
2-29) J. Klerer, J. electrochem. Soc. **108**, 1070-1071, 1961.
2-30) J. Klerer, J. electrochem. Soc. **112**, 503-506, 1965.
2-31) E. L. Jordan, J. electrochem. Soc. **108**, 478-481, 1961.
2-32) S. W. Ing, R. E. Morrison and J. E. Sandor, J. electrochem. Soc. **109**, 221-226, 1962.
2-33) W. A. Pliskin and H. S. Lehman, J. electrochem. Soc. **112**, 1013-1019, 1965.
2-34) P. White, Vacuum **12**, 15-19, 1962.
2-35) I. L. Chu and G. A. Gruber, Trans. met. Soc. AIME **233**, 568-572, 1965.
2-36) P. F. Schmidt and W. Michel, J. electrochem. Soc. **104**, 230-236, 1957.
2-37) P. F. Schmidt and A. E. Owen, J. electrochem. Soc. **111**, 682-688, 1964.
2-38) P. F. Schmidt, T. W. O'Keeffe, J. Droschnik and A. E. Owen, J. electrochem. Soc. **112**, 800-807, 1965.
2-39) P. F. Schmidt and J. E. Sandor, Trans. met. Soc. AIME **233**, 517-523, 1965.
2-40) J. R. Ligenza, J. appl. Phys. **36**, 2703-2709, 1965.
2-41) F. Singer, Z. Elektrochem. **32**, 382-395, 1926.
2-42) H. Edagawa, Y. Morita, S. Maekawa and Y. Inuishi, Jap. J. appl. Phys. **2**, 765-775, 1963.
2-43) W. A. Pliskin and P. P. Esch, J. appl. Phys. **36**, 2011-2013, 1965.

2-44) E. Kooi and M. M. J. Schuurmans, Philips Res. Repts **20**, 315-319, 1965.
2-45) T. E. Burgess and F. M. Fowkes, Spring Meeting Electrochemical Society, Cleveland, Ohio, Abstract 55.
2-46) M. V. Whelan, private communication.
P. V. Gray and D. M. Brown, Appl.Phys. Letters **8**, 31-33, 1966.
2-47) L. L. Rosier, IEEE Trans. **ED-13**, 260-267, 1966.
2-48) C. T. Sah and F. H. Hielscher, Phys. Rev. Letters **17**, 956-957, 1966.
2-49) M. M. Atalla, A. R. Bray and R. Lindner, Proc. int. elec. Engrs (London) Pt B. Suppl. **106**, 1130, 1959.
2-50) W. Shockley, H. J. Queisser and W. W. Hooper, Phys. Rev. Letters **11**, 489-490, 1963.
2-51) J. R. Matthews, W. A. Griffin and K. H. Olton, J. electrochem. Soc. **112**, 899-902, 1965.
2-52) E. H. Snow, A. S. Grove, B. E. Deal and C. T. Sah, J. appl. Phys. **36**, 1664-1673, 1965.
2-53) S. R. Hofstein, IEEE Trans. **ED-13**, 222-237, 1966.
2-54) Y. Matsukura and Y. Miura, Jap. J. appl. Phys. **5**, 180-182, 1966.
2-55) A. Goetzberger, Proc. IEEE **54**, 1454, 1966.
2-56) B. E. Deal, M. Sklar, A. S. Grove and E. H. Snow, J. electrochem. Soc. **114**, 266-274, 1967.
2-57) D. R. Kerr, J. S. Logan, I. J. Burkhardt and W. A. Pliskin, IBM J. Res. Dev. **8**, 376-384, 1964.
2-58) E. H. Snow and B. E. Deal, J. electrochem. Soc. **113**, 263-269, 1966.

3. DIFFUSION OF PHOSPHORUS INTO SILICON AND THE MASKING ACTION OF SILICON-DIOXIDE FILMS *)

Abstract

A neutron-activation analysis was used to study the behaviour of phosphorus during its diffusion into silicon and silicon-dioxide films. Certain correlations were found between the compositions of the oxide layers formed during the diffusions and the existing phase diagram of the system SiO_2-P_2O_5. In several experiments the concentration of phosphorus in silicon had reached the solubility limit, and a layer with very high phosphorus content was found to be present between the oxide film and the silicon substrate.

3.1. Introduction

It has been shown by several authors [3-1-3]) that diffusion of phosphorus in silicon cannot be described by a single diffusion coefficient. Tannenbaum [3-2]) and more recently Maekawa [3-3]) found a difference between the actual diffusion patterns and those obtained from resistivity measurements. The results can be explained in terms of an extra fast diffusion mechanism at high phosphorus concentrations. This may be due to diffusion of interstitial phosphorus atoms. Their presence may also affect the electron mobility in diffused layers. Both Tannenbaum and Maekawa found that phosphorus-doped silicon had a lower limit of resistivity of about 0·00035 Ωcm.

These conclusions are in agreement with the results of our experiments in which the diffusion of phosphorus into silicon and the masking effects of SiO_2 layers were studied by a neutron-activation analysis. P_2O_5 was applied as a source for the diffusion. Various P_2O_5 pressures were used, and the temperature of the silicon was varied from 920 to 1310 °C.

The phosphorus which diffuses into the silicon can be supposed to be formed according to the reaction

$$2\,P_2O_5 + 5\,Si \rightarrow 4\,P + 5\,SiO_2. \tag{3.1}$$

A certain amount of oxygen will also diffuse into the silicon, but because the solubility is low compared to that of phosphorus, this will be neglected. The oxide layer which forms at the surface is not a pure SiO_2 layer, because it will react with phosphorus oxide from the vapour. Assuming that the average composition of the oxide can be given by $y\,SiO_2.P_2O_5$, the value of y could be determined by activation analysis. When the diffusions are carried out in an oxygen ambient, the picture is somewhat more complicated because more SiO_2 will form than corresponds with reaction (3.1).

Reaction of P_2O_5 with SiO_2 can also be supposed to occur when the silicon

*) Published: J. electrochem. Soc. **111**, 1383-1387, 1964.

surface has been provided with an SiO_2 layer before the diffusion starts. Diffusion of phosphorus (oxide) through SiO_2 films has been studied by Sah et al. [3–4]) and Allen et al.[3–5]). In both cases *p*-type silicon underneath the oxide films was used as a phosphorus detector to check whether the masking conditions were fulfilled. Sah et al. concluded that a glassy layer of unknown composition was built up at the top of the oxide films and that a sharp boundary existed between the glassy layer and the oxide. This conclusion was confirmed by our analysis in which the phosphorus distribution in the oxide layers was measured directly.

In several cases the compositions of the oxide layers could be correlated with the phase diagram of the system SiO_2-P_2O_5 as given by Tien and Hummel [3–6]). This was especially the case in two stage processes, both for oxide layers formed during diffusion in silicon and formed at the top of a masking SiO_2 film.

If the diffusion of phosphorus in silicon is not fast enough to carry off all the phosphorus formed by reduction of phosphorus oxide, the phosphorus concentration in silicon will reach a saturation value and a new phase may form. In several cases we found indeed a layer with very high phosphorus concentration just at the interface between the silicon and the oxide layer formed during the diffusion. This enabled us to determine the solid solubility of P in Si.

3.2. Experimental procedure

Silicon wafers were cut perpendicular to the ⟨111⟩ direction of a float-zone single crystal of 5-Ωcm *p*-type silicon. After lapping with fine alundum powder, the slices were etched in a mixture of 2 parts HF (50%) and 5 parts HNO_3 (65%), so that 60 μ were removed from each side. The final thickness of the slices was about 250 μ. The diffusions were carried out in a two-zone furnace, supplied with a quartz tube of 2·5 cm diameter (fig. 3.1). A quartz boat filled

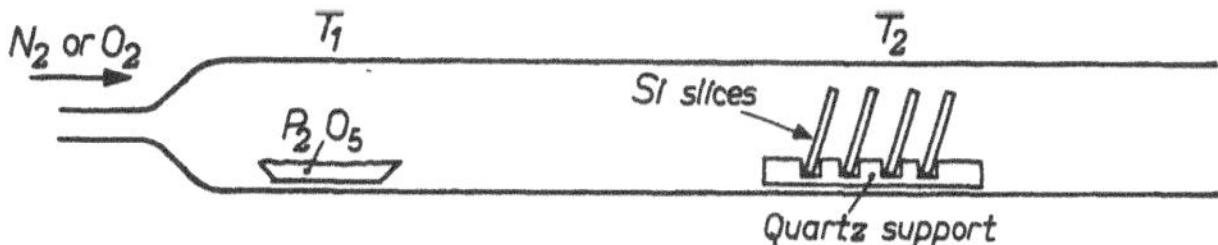

Fig. 3.1. Open-tube method for the diffusion of phosphorus into silicon.

with P_2O_5 was placed in the low-temperature zone (either 210 or 300 °C). The temperature of the silicon slices was varied between 920 and 1310 °C. Dry nitrogen or oxygen was used as carrier gas (gas flow 0·2 l/min). Heating of the silicon slices occurred by pushing them into the hot furnace, cooling by pulling them into a cold region of the quartz tube.

In a number of cases the diffusions were carried out in two stages: a prediffusion at 920 °C, followed by diffusion at a higher temperature in a second quartz tube, the walls of which had been exposed previously to P_2O_5 vapour at the diffusion temperature. However, before the silicon slices were inserted, the tubes were heated for a few hours at this temperature, without supply of P_2O_5. Under these conditions the amount of phosphorus per silicon slice did not change noticeably during the heat treatments.

After the diffusion the silicon slices together with a phosphorus standard (about 40 mg $(NH_4)_2HPO_4$) were irradiated in a neutron flux of 5.10^{11} neutrons/cm^2/s for a period of 5-10 days *). Except activity due to P^{32} (half-life 14·3 days), we found always some activity due to Au^{198} (half-life 2·7 days), which had also been formed during the neutron-activation process. This gold had been introduced during handling, etching, washing, and subsequent heat treatment of the slices. The also formed Si^{31} has a relatively short half-life (2·6 h) and decays practically completely in a few days.

If P^{32} as well as Au^{198} activity was present, the measurements were done with an NaI(Tl) well-type scintillation detector coupled with a gamma spectrometer. In those cases we measured both the distribution of phosphorus and of gold in the silicon and in the oxide film. The gamma-radiation spectrum of Au^{198} with a peak at 412 keV is superposed on the bremsstrahlungsspectrum of P^{32} (fig. 3.2). The amount present of both Au and P was determined by counting in two channels A and B according to the method of Elleman et al. [3-7]). Comparisons were made with irradiated standards. The conditions of

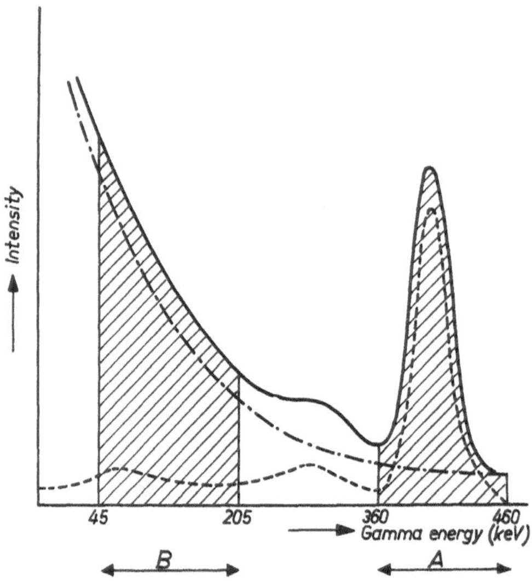

Fig. 3.2. Gamma spectrum of an irradiated silicon sample containing both phosphorus and gold (recorded after decay of Si^{31}). Dashed curve: Au^{198}; dashed-dotted curve: P^{32}; drawn curve: total.

*) For the irradiations we had the help of the Centre d'Etudes de l'Energie Nucléaire at Mol in Belgium. The samples were irradiated in the reactor BR I.

measurement for samples and standards were made as alike as possible by pipetting some of the standard solution onto non-irradiated silicon slices of the same thickness as the samples.

In the cases in which we were not interested in the behaviour of gold, we waited until the Au^{198} had disintegrated to a very low amount and measured the activity simply with a Geiger-Müller counter. The lowest amount of phosphorus which could be detected in this way was 10^{13} atoms.

3.3. Composition of oxide films formed during diffusion of phosphorus into silicon

3.3.1. *One-stage processes*

The activity of the silicon slices was counted before and after removal of the oxide layer (by a quick dip in an aqueous HF solution). The average composition of the layers could then be calculated by correlating the weigh decrease with the decrease in activity.

If no oxygen was present, a second method could be used. After the oxide layer had been removed, the residual activity gave the number of reduced P atoms and so, according to eq. (3.1), also the number of SiO_2 molecules in the oxide layers. Both methods gave the same results within 10%.

The oxide layers formed during the diffusion did not always show a "glassy" appearance. In some cases a crystalline structure was present; in other cases only very minor structure differences could be seen. According to the phase diagram of the system SiO_2-P_2O_5 given by Tien and Hummel [3-6]), compounds with compositions $SiO_2.P_2O_5$ and $2SiO_2.P_2O_5$ exist (fig. 3.3). Whether one or both of these compounds were present in the oxide layers could not be ascertained, but the presence of crystallites certainly occurred more frequently when the amount of P_2O_5 in the oxide layers was higher.

Using diffusion temperatures of 920, 1050, 1180 and 1310 °C, a P_2O_5 temperature of 210 °C, nitrogen as carrier gas, and diffusion periods of 15 and 60 min, the compositions of the oxide layers varied from $2SiO_2.P_2O_5$ to $4SiO_2.P_2O_5$. When the source was held at 300 °C, the P_2O_5 content was sometimes higher, but effects of non-constant P_2O_5 pressures due to aging of the source were of more influence in this case. It was found that a new source of P_2O_5 held at 210 °C could give a larger amount of deposited phosphorus than a source held at 300 °C, which had already been heated for a few hours.

When oxygen was used as carrier gas instead of nitrogen, the phosphorus contents of the oxide layers were lower, especially when the temperature of the silicon was relatively low and the diffusion time short. With a diffusion period of 15 min, the average oxide composition ranged from $10SiO_2.P_2O_5$ for experiments at 920 °C to $4SiO_2.P_2O_5$ at 1310 °C. For diffusion times of 60 min the difference between oxygen and nitrogen ambients became much less pronounced. Variations between $2SiO_2.P_2O_5$ and $4SiO_2.P_2O_5$ were found, depending on the

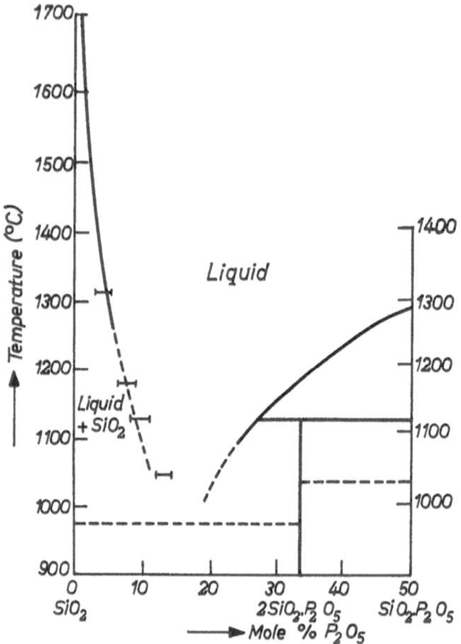

Fig. 3.3. Saturation values of oxide-layer compositions in two-stage diffusion processes, plotted in the phase diagram of SiO_2-P_2O_5.

state of the source. The large effect of the presence of oxygen in the short diffusion runs is explained by the fact that the oxidation velocity of silicon depends on diffusion through the oxide layer, which grows thicker during the experiment.

3.3.2. *Two-stage processes*

The prediffusion was done at 920 °C for 15 or 60 min in a nitrogen ambient. In the second stage the silicon slices were heated separately at a higher temperature for varying periods, using either a nitrogen or oxygen ambient. During the first stage oxide layers were formed with a composition of about $2SiO_2.P_2O_5$. During the second stage their compositions altered, owing to further reaction of P_2O_5 and Si resulting in SiO_2 and P. The changes of the oxide-layer compositions as a function of time have been given in fig. 3.4*a* for a diffusion in N_2 at 1050 °C. In the beginning the phosphorus content decreases rapidly, but then appears to become more or less constant at a lower limit ($7SiO_2.P_2O_5$). This means also that the amount of reduced phosphorus reaches a nearly constant value after some time (fig. 3.4*b*). The limiting composition of the oxide layers appeared to vary with temperature. Less phosphorus remained present when the temperature was higher. The values found at 1060, 1120, 1180 and 1310 °C have been plotted in fig. 3.3. They appear to coincide more or less with the solubility of SiO_2 in the phosphorus-containing liquid, as given in the phase diagram of Tien and Hummel.

A possible explanation can be obtained if one assumes that the reduction of

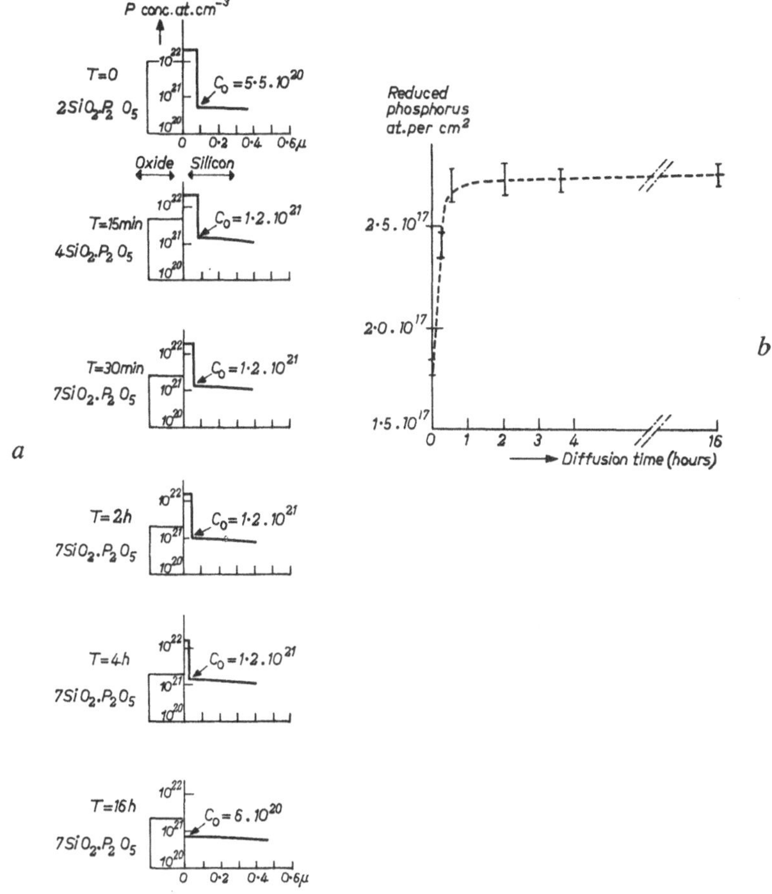

Fig. 3.4. Behaviour of P concentrations in oxide layers, silicon and interface layer during the second stage of a two-stage process, consisting of a prediffusion at 920 °C for 30 min followed by a diffusion at 1050 °C for a variable time T.
(*a*) The oxide layer obtains a constant composition, the surface concentration of P in Si remains constant as long as the interface layer is present.
(*b*) The total amount of phosphorus in silicon and interface layer goes to a constant value.

phosphorus oxide occurs at the interface between the silicon and the oxide layer. As the reduction of phosphorus oxide goes together with oxidation of silicon, one can understand that the reaction stops more or less as soon as a thin SiO_2 layer forms at the interface. This will occur when the oxide layer has reached such a composition that SiO_2 can no longer remain dissolved. The presence of an SiO_2 layer between the phosphorus-containing oxide layer and silicon could be shown clearly in the case where the second stage of the diffusion process was carried out in oxygen instead of nitrogen. By removal of successive layers of the oxide films, we could demonstrate the presence of an SiO_2 film between the phosphorus-containing oxide layer and the silicon substrate. This SiO_2 film may have contained some phosphorus, but we could not detect any, which means that it contained less than 10^{19} phosphorus atoms/cm^3. These oxidation experiments prove also that the species diffusing through the (phosphorus-containing) oxide layers was oxygen and not silicon.

3.4. Solubility limits and diffusion profiles of P in Si

If at a certain temperature the concentration of P in Si reaches the maximum solubility, one would expect the formation of a second phase of the diagram Si-P. In several cases we found indeed that after removal of the oxide layer by a quick dip in an HF solution, a thin layer with a high P content remained present at the surface. In our experiments this layer was always thinner than 0·1 μ. It could be removed by heating the slices in hot water or acid solutions, for example, immersion in an aqueous HF solution for several minutes. It was also possible to remove successive parts of the layer by anodic oxidation in a solution of KNO_3 in N-methyl acetamide and subsequent rinsing in an HF solution. The concentration of phosphorus in these layers was between 2.10^{22} and 4.10^{22} atoms/cm^3 in experiments below 1100 °C. In diffusion experiments at 1180 and 1310 °C the concentration appeared to be somewhat lower, although still of the order of 10^{22} atoms/cm^3. This is in agreement with the phase diagram of the system Si-P as given by Giessen and Vogel [3-8]), partly given in fig. 3.5. In this system a compound SiP can be in equilibrium with silicon

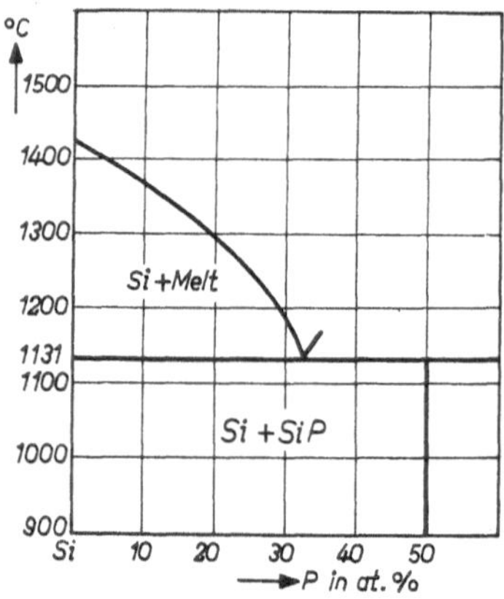

Fig. 3.5. The system Si-P (according to Giessen and Vogel [3-8])).

saturated with phosphorus, as long as the temperature is below 1131 °C. Above that temperature equilibrium can exist between silicon saturated with phosphorus and a melt with a rather high phosphorus content. It seems probable that the interface layers which were found in several of our experiments are due to formation of one of these phases, although some oxygen may have been incorporated. When an "Si-P" phase was present at the surfaces, the surface concentration of the phosphorus diffusion was found to be determined only by the temperature of the heat treatment. Therefore we were able to determine the solubility limits of phosphorus in silicon for a few temperatures. Results are given in fig. 3.6. Our results show a higher solubility of phosphorus in silicon

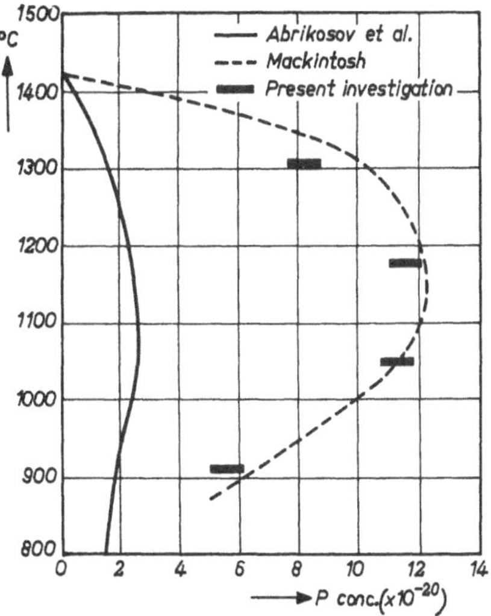

Fig. 3.6. Solubility limits of P in Si.

than reported by Abrikosov et al.[3-9]), whose values were obtained from microhardness measurements in the system Si-P. The results are very close to the lower limits of solubility found by Mackintosh [3-10]). This agreement is rather fortuitous, however, because Mackintosh's results were obtained from measurements of sheet resistance and *pn*-junction depth.

In a way similar to that described by Tannenbaum [3-2]) and Maekawa [3-3]) we measured both the diffusion profiles and the distribution of electrical conductivity in the diffused layers. Good agreement with their measurements was found, such as a minimum in resistivity of 3.10^{-4} Ωcm for phosphorus-doped silicon. We found that this minimum corresponded to a phosphorus concentration of $6{\cdot}5.10^{20}$ per cm^3. The minimum in resistivity did not necessarily occur at the silicon surface, as the solubility of phosphorus in silicon can be higher.

The surface concentration of phosphorus found in a two-stage process is often lower than corresponds with the solid solubility. However, as long as a silicon-phosphide phase is present at the surface, the surface concentration of phosphorus will have its maximum value at the given temperature. As an example of the behaviour of P concentration in the oxide, interface layer and silicon in a two-stage process, a schematic picture has been given in fig. 3.4*a*. In this case both diffusion steps were carried out in nitrogen; in the first step the silicon was held at 920 °C for 30 min with P_2O_5 at 220 °C and in the second step at 1050 °C for various times.

Formation of surface films with phosphorus concentrations as high as 10^{22} cm^{-3} has been reported neither by Tannenbaum nor by Maekawa. Also in our experiments such a layer was not always found to be present. Obviously the experimental conditions play an important role. We found that the presence of these surface layers was less pronounced if the P_2O_5 content of the oxide layers was decreased and the temperature of the silicon increased. This can be understood if one considers that the formation of an Si-P phase will be enhanced when more P_2O_5 can be reduced and when the diffusion into silicon is not fast enough to carry away the reduced phosphorus atoms from the surface. As discussed before, the P_2O_5 content of the oxide layers depends on many factors such as the presence of oxygen. It may well be that in Tannenbaum's experiments the highly doped layers were not found because the silicon was preheated at the diffusion temperature for 10 min in oxygen before admitting the P_2O_5 (private communication). In Maekawa's experiments the source was $H_4P_2O_7$, so that comparison is more difficult.

3.5. The masking action of SiO_2 films against P_2O_5 diffusions

Some study on this subject was done using thermally grown SiO_2 layers as a mask. These oxide films had been made by heating the silicon slices during 16 hours at 1200 °C in wet oxygen (oxygen bubbled through water at 30 °C), resulting in an oxide thickness of 1·1 micron. The oxidized slices were then subjected to phosphorus diffusion like described in sec. 3.2. To get an impression of the distribution of phosphorus in the oxide layers, the slices were irradiated by neutrons and the decrease of P^{32} activity was measured after successive removal of sheets of the oxide film in a diluted solution of hydrofluoric acid.

In each case the activity disappeared after removal of a part of the oxide film, pointing to a sharp boundary between a phosphorus-containing top layer and the remaining part of the oxide film. In two-stage processes (no P_2O_5 offered during the second stage) again an equilibrium composition of the glassy layer was found, if the diffusion period was sufficiently long. In a case in which the second stage of the process was carried out at 1120 °C, the final glass composition was calculated to be between $8SiO_2.P_2O_5$ and $12SiO_2.P_2O_5$, again showing a reasonable correspondence with the solubility of SiO_2 in the glassy phase at this temperature.

There will probably be some diffusion from the phosphorus-containing top layer into the underlying "pure"-SiO_2 film, but this could not be detected. It is felt that this diffusion has not much influence on the masking action of the SiO_2 film: masking will be effective until the phosphate glass has completely penetrated through the SiO_2 film. We heated *p*-type samples covered by an SiO_2 layer with phosphate glass on top for several days at 1100 °C in nitrogen and did not notice any phosphorus diffusion into the silicon when no extra P_2O_5 was offered during the heating. SiO_2 films are thus most effective in masking in the described two-stage processes.

3.6. The use of phosphorus diffusions in planar processes

Phosphorus diffusions are used frequently for making planar structures, i.e. structures in which the *pn* junctions formed during diffusion are protected by the same oxide film, which was used for masking. Figure 3.7 shows planar-diode structures assumed to be formed by local diffusion of phosphorus into *p*-type

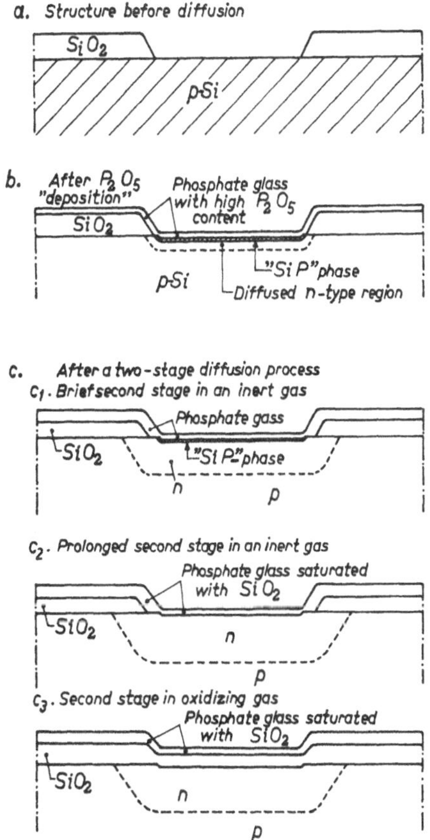

Fig. 3.7. Diffusion of phosphorus in a "planar" process. During P_2O_5 deposition an "SiP" phase may form at the boundary of the phosphate glass and the silicon when the P_2O_5 concentration is high. When during the second stage no more P_2O_5 is offered, the SiP phase tends to disappear and the phosphate glass, both on top of the oxide and in the window, tends to become saturated with SiO_2.

silicon through a window in an SiO_2 film. In semiconductor-device technology P_2O_5 is often made in the diffusion tube by a reaction of P or compounds such as $POCl_3$ or PH_3 with oxygen. This prevents much troubles due to a non-constant vapour pressure of P_2O_5 as a consequence of aging of the P_2O_5 source. This phenomenon, which was mentioned earlier in sec. 3.3.1, is probably due to a change in the structure of the P_2O_5. Traces of water may play an important role in this effect.

As described in the previous sections, phosphorus diffusion with the aid of P_2O_5 will result in formation of a phosphorus-containing oxide, both in the area where the silicon is not covered by SiO_2 and in the top of a masking SiO_2 film- When P_2O_5 is offered only during a part of the diffusion time the phosphorus. containing glassy layers in both instances will tend to have the composition given by the solubility limit of SiO_2 in the glass at the diffusion temperature (fig. 3.7 c_2). As was discussed in sec. 3.3.2, the n-type region formed by the phosphorus diffusion may also become separated from the phosphate glass by a "pure"-SiO_2 film when the diffusion is carried out in oxygen (fig. 3.7 c_3).

3.7. Gettering of metallic impurities by surface films

In our samples nearly always some activity of Au^{198} was present after irradiation and therefore we were able to study the behaviour of small amounts of gold during oxidation and diffusion treatments. After oxidation (wet O_2, 1200 °C) we detected an accumulation of gold at the Si-SiO_2 interface varying from 10^{10} to 10^{13} atoms per cm^2. In other experiments we found similar amounts of copper. There are indications [3-11]) that the major part of these metallic impurities is not distributed homogeneously, but rather is present in local precipitates at the Si-SiO_2 interface. As such they may cause soft reverse I-V characteristics of planar pn junctions.

It has been found [3-12,13]) that the properties of pn junctions can often be improved by a gettering treatment, i.e. a treatment of the silicon in contact with a material in which the metal impurities tend to be dissolved. Phosphate glass is often considered as such a material. We found indeed that the gold concentration in the silicon and at the Si-SiO_2 interface became often very low after a P_2O_5 diffusion. The gold appeared to be present in the phosphate glass and in the SiP phase. Gettering appeared to be improved by a high P_2O_5 content of the glassy layer and a high temperature. Below about 1100 °C gettering of gold appeared to be not very effective.

In chapter 7 the gettering effects of a phosphate glass for alkali ions in silicon-oxide films will be considered. This effect is sometimes of large importance for improvement of the surface properties of oxidized silicon.

Note Since this work was presented first (Meeting of the Electrochemical Society, Pittsburgh, April 1963) various authors [3-13-15]) have shown that precipitates may exist at the surface of a silicon crystal after this has been subjected to phosphorus diffusion. Precipitates may also form in the diffused region during cooling. In some cases they could be shown to consist of SiP [3-13]). A number of compositions of phosphate-glass layers grown under various conditions were verified by Snow and Deal [3-16]) using gravimetrical methods.

REFERENCES

3-1) V. K. Subashiev, A. P. Landsman and A. A. Kukharskii, Sov. Phys. solid State **2**, 2406-2411, 1961.

3-2) E. Tannenbaum, Solid State Electronics **2**, 123-231, 1961.

3-3) S. Maekawa, J. phys. Soc. Japan **17**, 1592-1597, 1962.

3-4) C. T. Sah, H. Sello and D. A. Tremere, J. Phys. Chem. Solids **11**, 288-298, 1959.

3-5) R. B. Allen, H. Bernstein and A. D. Kurtz, J. appl. Phys. **31**, 334-337, 1960.

3-6) T. Y. Tien and F. A. Hummel, J. Am. ceram. Soc. **45**, 422-424, 1962.

3-7) T. S. Elleman, J. E. Howes Jr and D. N. Sunderman, Int. J. appl. Rad. Isotopes **12**, 142-152, 1961.

3-8) B. Giessen and R. Vogel, Z. Metallkunde **5**, 174-177, 1959.

3-9) N. K. Abrikosov, V. H. Glazov and L. Chên-Yiian, Russ. J. inorg. Chem. **7**, 429-431, 1962.

3-10) F. A. Trumbore, Bell Sys. techn. J. **39**, 205-233, 1960.

3-11) M. M. Atalla, Properties of elemental and compound semiconductors (Met. Soc. Conf.), Interscience Publishers, New-York-London, 1959, Vol. 5, pp. 163-181.

3-12) A. Goetzberger and W. Shockley, J. appl. Phys. **31**, 1821-1824, 1960.

3-13) P. F. Schmidt and R. Stickler, J. electrochem. Soc. **111**, 1188-1189, 1964.

3-14) M. L. Joshi, J. electrochem. Soc. **113**, 45-48, 1966.

3-15) R. Gereth and G. H. Schwuttke, Appl. Phys. Letters **8**, 55-57, 1966.

3-16) E. H. Snow and B. E. Deal, J. electrochem. Soc. **113**, 263-269, 1966.

4. EFFECTS OF LOW-TEMPERATURE HEAT TREATMENTS ON THE SURFACE PROPERTIES OF OXIDIZED SILICON *)

Abstract

Oxidized silicon slices were subjected to heat treatments in various environments in the range 300-500 °C either with or without an aluminium electrode present on top of the oxide. Several results point to a large effect of hydrogen in these heat treatments, after which the surfaces often exhibited an increased *n*-type character. These effects are primarily due to the disappearance of interface states, although in some cases charge redistributions in the oxide-silicon system also play a significant role.

4.1. Introduction

In chapter 1 it was discussed how capacitance measurements on metal-oxide-silicon structures can give information about the charge distributions at an oxidized silicon surface. Certain relations were shown to exist between high-frequency capacitance measurements, carried out by Whelan [4-1]), and MOS-transistor characteristics. The presence of electron-trapping centres at the Si-SiO_2 interface as well as fixed charge in the oxide has an influence on these electrical characteristics. In this paper we will further relate capacitance and channel-current measurements on MOS-transistor structures to the changes in the surface properties of oxidized silicon which may occur during heat treatment at relatively low temperatures (300-500 °C). We studied the effect of the ambient gas in such treatments and also the influence of an aluminium electrode if present on top of the oxide during heating. It is known that heating in the presence of such an electrode may cause formation of *n*-type surfaces [4-2-5]).

Presence of hydrogen and water vapour during heat treatment often causes similar effects [4-5-8]). Although the *n* shifts have been explained in a few cases as being due to formation of donor-type surface states, other experiments [4-4,6,9]) indicate that annihilation of electron-trapping centres plays a more important role. Balk [4-9]) has proposed that the effect of hydrogen during annealing is due to a chemical saturation with H atoms of certain unsaturated bonds at the oxide-silicon interface. The experiments to be described in this paper suggest that the main effect of water and an aluminium electrode on the oxide during low-temperature heat treatments has to be explained in a similar way, although charge redistributions in the oxide-silicon system cannot always be neglected.

In the experiments we generally made use of oxide films which were covered by a phosphate glass, i.e. the same type as Whelan [4-1]) used in some capacitance versus voltage measurements. The large number of active surface states in these cases made it convenient to study the influence of heat treatment on the electrical characteristics of the MOS structures.

*) Published: Philips Res. Repts **20**, 578-594, 1965.

4.2. The MOS transistor

4.2.1. *Structure and method of preparation*

The circular geometry of the MOS devices employed is shown in fig. 4.1. The structure was made by local phosphorus diffusion into *p*-type material. Slices, cut perpendicular to the ⟨111⟩ axis of a 5-Ωcm indium-doped silicon crystal,

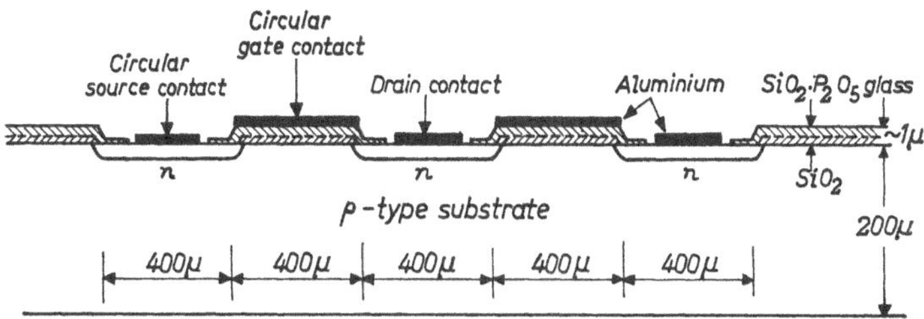

Fig. 4.1. A cross-section of the MOS-transistor structure used in several experiments.

were first lapped with fine alundum powder, and then etched in an HF-HNO_3 mixture so that a layer of about 60 microns was removed from each side. Thermal oxidation was carried out at 1200 °C for 16 hours in wet oxygen, made by bubbling oxygen through water at 30 °C. Photo-etching techniques were used to make windows in the oxide film with the view of producing *n*-type "source" and "drain" regions. The phosphorus diffusion was performed in nitrogen, employing a source of P_2O_5. The silicon was first heated for 30 minutes at 920 °C in a two-zone diffusion system, in which the P_2O_5 source was held at 220 °C. The source was then cooled to room temperature and the silicon slices were heated at 1150 °C for 4 hours. The samples were cooled rather slowly by turning off the power supply to the furnace. The *pn*-junction depth obtained after this procedure was 6 microns. The total oxide thickness at this stage was about 1·2 micron. The top layer (about 0·5 micron) of an oxide film made in this way consisted of a mixed oxide of phosphorus and silicon with a composition of about 12 $SiO_2.P_2O_5$ [4–10]). The contacts indicated in fig. 4.1 were made by vapour deposition of aluminium followed by photo-etching. During the evaporation process the silicon was not heated.

4.2.2. *Interpretation of electrical characteristics*

Information about surface properties of oxidized silicon was obtained by measuring the current between source and drain contact of the MOS transistors as a function of the gate voltage. The drain region was made positive in these measurements and thus reversely biased with respect to the bulk. In general we measured the channel current (I_D) at a drain voltage (V_D) which was larger than the pinch-off voltage, thus in the saturated region of the I_D-V_D characteristics. In fig. 4.2 an example of these characteristics has been given. An essential part of the preparation of this sample was a heat treatment in water vapour at the end of the phosphorus diffusion. Without such a treatment much higher gate voltages would have been needed to induce the same drain currents. In sec. 1.7 it was shown that measurements of the capacitance of the MOS structure as a function of gate voltage suggest that this is due to the presence of surface states in which the induced negative charge is trapped instead of being mobile.

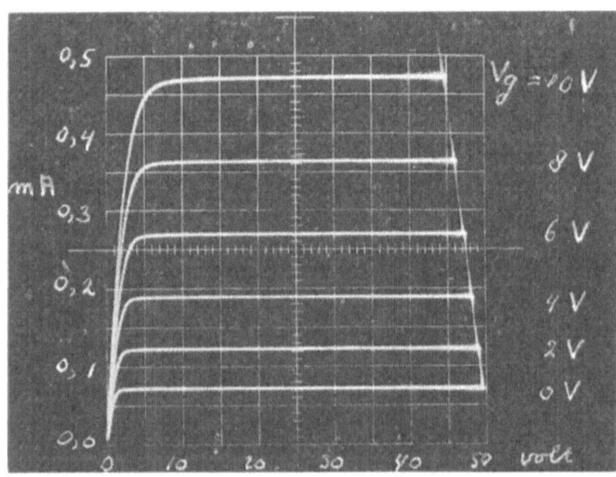

Fig. 4.2. I_D-V_D characteristics of a MOS transistor with varying gate voltage. This sample had been heated during 30 minutes at 450 °C in wet nitrogen before the aluminium electrodes had been deposited.

In the work presented in this chapter in addition to I_D-V_G measurements high-frequency (500 kc/s) capacitance measurements on the MOS structures were used also to study the effect of low-temperature heat treatments on oxidized surfaces. The capacitance measurements were primarily used to obtain the voltage that was necessary to obtain "zero band bending" at the silicon surface. In many cases this value will be given in tabular form together with the voltages which are necessary to induce various channel currents I_D. The difference ΔV_G between the gate voltage which is needed to induce a drain current of 10 μA and the voltage which has to be applied for zero band bending is used as an indication of the number of electron-trapping centres at the surface. The lowest value of ΔV_G which we obtained was about 8 volts. It can be calculated that a voltage difference of this order of magnitude can be expected when no surface states are present and a reasonable value of the electron mobility in the channels is assumed. We found, however, that in some cases ΔV_G was even larger than 100 volts. The values of these voltage differences are not a quantitative measure for the number of surface states (electron-trapping centres), because the mobility of the electrons in the channels is not known, and is presumably also affected by the presence of these centres. Moreover, only a part of the surface states will exchange charge with silicon when the band bending is varied from zero to the value which corresponds to a channel current of 10 μA. Nevertheless, we think that the large variations in the indicated voltage difference ΔV_G point to large variations in the surface-state density.

4.3. Effect of ambient gas during heat treatments

Table 4-I shows the effect of ambients for heat treatments of 30 minutes at 450 °C. Wet oxygen or nitrogen was made by bubbling the gas through water at room temperature. These heat treatments were carried out directly at the end of the phosphorus diffusion. After the heat treatments photo-etching and vapour deposition were employed to make aluminium contacts to source, drain, gate and *p*-type substrate. A rough idea about the variation of surface charge with bias can be obtained from table 4-I and the following tables by applying the formula $\Delta Q = C_{ox} \Delta V$. As the oxide capacitance C_{ox} was about $2{\cdot}7.10^{-9}$ F/cm², a voltage difference of 1 volt corresponds to a change in surface charge of about $1{\cdot}7.10^{10}$ unit charges per cm².

TABLE 4-I

Effect of ambient gas during heat treatment of oxidized silicon for 30 min at 450 °C. The gate voltages V_G necessary for flat-band conditions ($y_s = 0$) were obtained from C-V_G measurements on MOS-transistor structures, which were also used to measure the gate voltage necessary to induce various values of the drain current

	V_G (in volts) necessary to induce:						ΔV_G (V_G at $I_D = 10$ μA minus V_G at $y_s = 0$)
	$y_s = 0$ (V_f)	$I_D =$ 1 μA	$I_D =$ 10 μA	$I_D =$ 100 μA	$I_D =$ 1 mA	$I_D =$ 2·5m A	
before treatment	–30	66	90	115	154	180	120
after treatment in:							
dry N_2	–29	70	97	121	153	180	126
dry O_2	–42	50	75	102	142	170	117
wet N_2	–12	–6	–4	2	19	34	8
wet O_2	–16	23	37	53	86	106	53
dry H_2	–16	–10	–8	–2	16	30	8

Obviously, heating in dry nitrogen did not have much effect. The main effect of dry oxygen seems to be that of a shift of both the C-V_G and the I_D-V_G characteristics to more negative or less positive gate voltages. This may be explained as an increase of the effect of fixed charge in the oxide. The number of active electron-trapping centres at the interface remains of the same order of magnitude.

As was also discussed by Whelan [4-1]), the effect of a treatment in wet nitrogen has to be explained as being primarily due to the disappearance of the active trapping centres. It is not well possible to conclude whether the fixed charge increases or decreases during the treatment (in chapter 7 it will be made plausible that V_G at "$y_s = 0$" (V_f) is an indication for the effective oxide charge; this would mean that the fixed charge has slightly decreased).

The effect of hydrogen did not differ very much from the effect of wet nitrogen. The surface states disappeared in both cases. The remaining surface charge was somewhat more positive than in the case of heat treatment in water vapour.

The disappearance of the surface states at the Si-SiO_2 interface during heating, both in hydrogen and in water, suggests that also in the latter case the effect is due to a reaction of the trapping centres at the interface with hydrogen. To cause such an effect, the hydrogen component of the water should diffuse through the oxide. One might suppose that H_2O can be chemically reduced in the top layer of the oxide film (the phosphate glass in these cases). Oxygen may then be built in in the oxide, whereas the hydrogen component of the water diffuses partly through the oxide film and reacts with the trapping centres, thus making them inactive. This assumption is supported by the results of the treatment in wet oxygen instead of in wet nitrogen (table 4-I). The presence of oxygen during heating diminished the influence of water vapour. This may be explained by assuming that the presence of oxygen makes reduction of water less easy or, in other words, that the hydrogen becomes partly trapped at the excess oxygen atoms instead of at the silicon surface, so that diffusion of hydrogen (or hydroxyl groups) into the oxide is slowed down.

We observed that during heating in dry nitrogen or oxygen a considerable number of trapping centres disappeared when the heat treatments were not carried out directly after the phosphorus diffusions, but instead after the samples had been stored in room air for several minutes. Presumably a sufficient amount of water is adsorbed in this time to have some influence during heat treatment.

We also found that channel formation during heat treatment in a wet ambient occurred much sooner when the phosphate glass was removed before heating. By combining this knowledge with measurements of C-V_G curves of MOS structures we concluded that this is partly due to a more rapid disappearance of the electron-trapping centres, but also that the amount of fixed surface charge was larger after heat treatment when the phosphate glass had been removed before heating. Obviously the phosphate glass has a masking action for the diffusion of water and perhaps other impurities into the oxide. This effect may be of importance in connection with the stabilizing action of a phosphate glass on planar structures [4-11]).

4.4. Influence of an aluminium electrode during heat treatment of MOS structures

It is known that during heat treatment of oxidized silicon samples in the range of 300 to 700 °C the silicon surface may acquire a considerably stronger n-type character if the metal can be supposed to react with SiO_2 [4-2-5]). One of these reactive metals is aluminium, which is often used for the metal contacts on planar structures. Formation of n-type surfaces may occur during contacting processes or life tests of such structures, especially in those cases where an aluminium electrode lies on top of the oxide film. The effect of such an electrode can often be increased when an electrical connection is made between the metal and the semiconductor. This has been explained [4-2]) by relating the phenomena to the charge redistributions which may occur when a bias is applied across a MOS system. When this external voltage is made zero, chemical-potential differences in the MOS system may cause similar phenomena.

On the other hand, however, it is known that the presence of an aluminium electrode on oxidized silicon during heating can cause the disappearance of active surface states, just as in the case of heat treatment in hydrogen or water vapour as discussed in the previous section. It has been proposed [4-7]) to use this method to increase the transconductance of MOS field-effect transistors. Inversion-layer formation on p-type silicon during heat treatment in the presence of an aluminium electrode may thus be related both to charge redistributions and to the disappearance of the influence of trapping centres at the surface.

As in our samples a large number of such centres were present at the end of the phosphorus-diffusion process, it was relatively easy to observe the effect of heat treatment (which was done in dry nitrogen) in the presence of an aluminium electrode on their disappearance. Again C-V_G and I_D-V_G measurements on MOS transistors with the dimensions given in fig. 4.1 were used to study these annihilation effects, which were found to be very fast at temperatures in the range of 400-500 °C. The trapping centres disappeared within a few minutes in these cases. As these centres are presumably present near the Si-SiO_2 interface, a fast-diffusing species should be encountered in this process, as the effects are due to the presence of aluminium at the top of the oxide. To answer the question whether the diffusing species was charged or neutral, we tried to influence the effect by applying a bias across the MOS structure during heat treatment. The results of heat treatments without bias and with either positive or negative bias on the metal electrode are shown in fig. 4.3. The increase in transconductance does not seem to be influenced by the bias, although its presence has certainly some effect on the I_D-V_G curve in the sense of a shift in the direction of less positive voltages. This effect can be explained as due to a charge redistribution

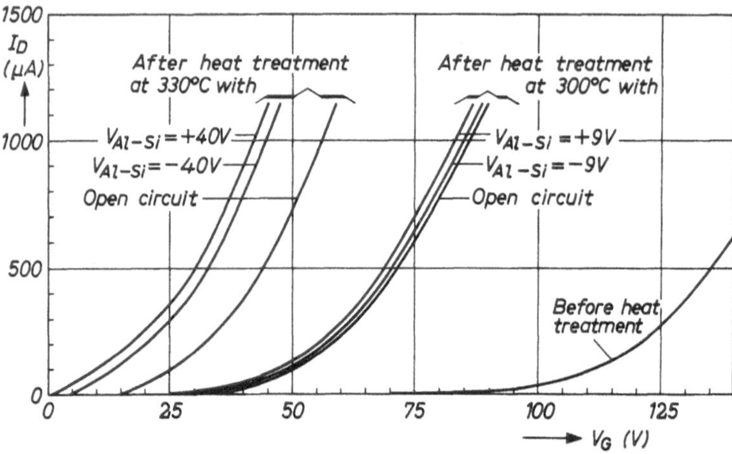

Fig. 4.3. Effect of a heat treatment on the I_D-V_G characteristics of MOS-transistor structures with and without an applied bias between the aluminium electrode and the silicon substrate. The indicated bias was applied during the whole heating cycle in which the samples were heated in 8 minutes from room temperature to the indicated temperature, which was maintained for 5 minutes, after which cooling to room temperature was carried out in less than 1 minute. The ambient gas during heat treatment was nitrogen.

in the metal-oxide-silicon system. The experiments suggest therefore that, although certain charged species may move during the heat treatment, the diffusing species that causes the disappearance of the surface states is neutral. Measurements performed by Whelan [4-1]) on the changes of the *C-V* curves of MOS structures due to heating under bias point in the same direction.

The effect of heat treatment in the presence of an aluminium electrode is strongly similar to the effect of hydrogen and water vapour during heat treatments, discussed in the previous section. In the cases of treatment in hydrogen or water it has been suggested that in both cases the trapping centres disappeared due to reaction with hydrogen. Therefore a possible explanation for the observed effect would be that during the reaction of Al and SiO_2 hydrogen is formed, which diffuses in a neutral form (molecules or atoms) through the oxide to the silicon surface. The influence of the aluminium electrode could then be considered as an enhancement of the effect of the presence of water during heat treatment, discussed in the previous section. This is caused by the fact that the aluminium acts as a reducing agent. Remember that the presence of oxygen had a retarding action on the effect of water on the annihilation of the surface states (table 4-I).

The formation of hydrogen due to a reaction of aluminium and oxide can be understood when the oxide contains hydrogen. We have found, however, that the considered influence of aluminium is present for oxides made in wet as well as in dry ambient. It may be, however, that even in our driest ambients sufficient water or hydrogen was present to form a certain amount of hydroxyl groups in the oxide. More likely, water adsorbed at the oxide or aluminium surface during handling between the end of the oxidation or the phosphorus-diffusion process and the following heat treatment plays an important role.

In table 4-II the action of an aluminium electrode has been compared with the action of hydrogen without the presence of an aluminium electrode. In both cases heat treatment was carried out at 300 °C for 10 minutes. In the case of

TABLE 4-II

Effects of heat treatment on MOS-transistor characteristics.
Sample A: no heat treatment.
Sample B: heat treated in H_2 at 300 °C during 10 min. No Al electrode present during heating.
Sample C: heat treated in N_2 at 300 °C with an Al electrode present upon the oxide.
The voltage needed to obtain zero band bending was deduced from capacitance measurements

	V_G (volts) at:		ΔV_G
	$y_s = 0$	$I_D = 10$ μA	$V_G(I_D = 10\ \mu A)$ $-V_G\ (y_s = 0)$
A	—30	91	121
B	—29	60	89
C	—22	26	48

the treatment in hydrogen the electrodes were applied after the treatment. It is clear from this table that the influence of an aluminium electrode on the disappearance of electron-trapping centres was faster at this temperature than the effect of hydrogen of 1 atm. As it can hardly be expected that due to the reaction of Al and water at the oxide surface more H_2 will form in the oxide than can be introduced in a hydrogen ambient, the comparison of these results leads to the conclusion that the effect of a reacting Al electrode is not simply due to formation of H_2, but that a more reactive species is encountered in this process. It is suggested that a chemical reaction of Al and the oxide or water at its surface results in the formation of hydrogen atoms which live long enough to penetrate through the oxide before they have been able to recombine with another H atom or to react somewhere in the oxide.

The influence of temperature on the effect of a reacting aluminium electrode on the disappearance of surface states is illustrated in table 4-III. The value of ΔV_G may again be considered as a measure for the number of these states, as it represents the difference between the gate voltage necessary to induce 10 μA channel current and the voltage to obtain flat-band conditions. In a case of "no" surface states (table 4-I, heat treatment in H_2 or H_2O vapour) ΔV_G is about 8 V. This value is also reached in a short time for the treatment at 400 °C with aluminium present, but not at the lower temperatures. Instead, the reaction seems to slow down after a few minutes. This may be explained by considering that during heating the aluminium electrode becomes covered by an aluminium-oxide film, which inhibits further reaction, especially at the lower temperatures.

TABLE 4-III

Influence of an aluminium electrode on the disappearance of electron-trapping centres at oxidized silicon surfaces during heat treatment in nitrogen at various temperatures; ΔV_G is the difference in gate voltage, necessary to induce a channel current of 10 μA and the voltage which had to be applied to obtain a MOS capacitance corresponding to zero band bending at the silicon surface. Before heat treatment ΔV_G was about 120 V

temperature of heat treatment (°C)	ΔV_G after various heating periods			
	10 min	30 min	60 min	90 min
300	28	20	18	17
350	15	11	10	10
400	11	8	8	8

It also appeared to be possible to retard the effect of a reactive aluminium electrode by heating the samples in an oxidizing ambient before the aluminium was deposited. As shown in table 4-I, heat treatment in dry oxygen had only a little influence on the number of active surface states. In table 4-IV the effect of an Al electrode on these states during heat treatment at 300 °C has been compared for various samples of which one was first treated in oxygen at 450 °C for 30 minutes. The effect of a pre-treatment in oxygen is analogous to the influence which oxygen showed during heating of oxide-covered samples in water vapour (table 4-I) and may be explained in a similar way. Hydrogen, formed at the oxide-aluminium interface may become trapped at the excess-oxygen sites instead of moving to the silicon surface. Another explanation could be that the presence of some excess oxygen causes an easier formation of an aluminium-oxide film, which retards further reaction.

Table 4-IV also gives results for heat treatments in wet nitrogen at 400 °C (at this temperature there was only a small influence on the number of surface states) and re-heating in dry nitrogen at 300 °C with an aluminium electrode present. In these cases the effect of the reactive electrode was diminished even more than in the case of a pre-heat treatment in oxygen. This result is somewhat surprising, because one would expect more formation of hydrogen in the reaction of the oxide with aluminium, i.e. an increased effect on the disappearance of surface states, when water is present in the oxide. But one should realize that the effect of heat treatment in water vapour has been explained in the previous section as an oxidation of the upper layer of the oxide, during which the evolved hydrogen diffuses partly through the oxide to combine with the active surface states. From this point of view it seems reasonable to assume that during heat treatment of the oxidized slices in a wet atmosphere fewer hydroxyl groups form in the upper layer of the oxide than correspond with the excess oxygen built in. The heat treatment in water vapour may thus have caused formation of defects in the oxide, in which the evolved hydrogen atoms may become trapped instead of reaching the silicon surface. The experiments indicate in any case that the interaction between Al, H_2O and SiO_2 is more complex than would be expected from a simple chemical point of view.

TABLE 4-IV

Influence of pre-heat treatment in oxygen or wet nitrogen on the effect of an aluminium gate electrode on the disappearance of surface states during heating; ΔV_G is the difference in the gate voltage to obtain a drain current of 10 μA and the gate voltage at zero band bending

		no pre-heat treatment	pre-treatment		
			30 min O_2 450 °C	10 min wet N_2 400 °C	30 min wet N_2 400 °C
ΔV_G (V) *before* and ΔV_G (V) *after* various heat treatments in presence of an Al electrode		120	115	116	107
	10 min 300 °C	28	55	60	94
	30 min 300 °C	20	40	40	88
	50 min 300 °C	18	37	32	83

4.5. Low-temperature treatments of silicon oxidized in dry oxygen

Up to now we have considered in this paper oxide films to which a phosphorus diffusion in nitrogen had been applied. The top layer of this oxide consisted thus of a mixed oxide of silicon and phosphorus. We have also considered oxide films made in dry or wet oxygen on structures similar to that in fig. 4.1, with the exception that the distance between the *n*-type source-drain regions was only 18 microns. In this process the phosphorus diffusion was also carried out in two stages, but after the first stage (30 min 1050 °C) the oxide layer was dissolved in an aqueous HF solution and a new film was regrown by thermal oxidation in dry or wet oxygen.

On the surfaces coated by steam-grown oxides we found a much lower number of surface states than in the cases of dry oxidation. In this section we will further discuss only experiments carried out after oxidation of the samples in dry oxygen at 1200 °C for 1 hour. The oxide thickness was 0·2 micron in these cases.

Apart from *npn* structures we considered also *pnp* structures made by boron diffusion into 1-Ωcm *n*-type material. The diffusion was again carried out in two stages: deposition of B_2O_3 for 5 minutes at 1050 °C, after which the oxide film was removed and a new oxide film was regrown in dry oxygen at 1200 °C for 1 hour. As regards the method of oxidation, the *npn* and *pnp* structures are thus comparable.

4.5.1. *npn structures*

Measurements of capacitance together with channel conductance versus gate voltage pointed to the presence of interface states, althoughless in number than in the case of oxides covered by a phosphate glass, considered in the previous sections. Low-temperature heat treatments in hydrogen, in water vapour or in the presence of an aluminium electrode could be used to decrease their number. Water vapour had already large effects at 400 °C, which was not so much the case in the samples considered in the previous sections (table 4-IV). The difference in behaviour between the various types of oxide suggests again (compare also the last paragraph of sec. 4.3) that a phosphate glass can have a masking action for the diffusion of water or components of it into the oxide, although the thickness of the oxide may also be of influence.

A pre-heat treatment in water vapour at 400 °C could therefore not be used either to slow down channel formation below an aluminium electrode during heating. However, at lower temperatures similar effects were observed as in the case of the oxide covered by phosphate glass, shown in table 4-IV. In fig. 4.4 a number of drain characteristics have been given before and after various heat treatments at 300 °C. The transconductance of the devices increased after heat treatment in dry nitrogen with an aluminium electrode present and also to a certain extent during heat treatment in wet nitrogen, when no electrode was present on the oxide. However, a second heat treatment at 300 °C for 10 minutes with an aluminium electrode present had then almost no effect on the I_D-V_G curves.

Using an oxidation of 1 hour in dry oxygen at 1200 °C and 1-10 Ωcm *p*-type silicon as starting material, an inversion layer was always found at the surface, when low-temperature heat treatments had been carried out to remove the electron-trapping centres at the surface. With suitable heat treatments a number of these centres can remain present so that, even after mounting of the device on a transistor header (heating is then carried out with aluminium present upon the

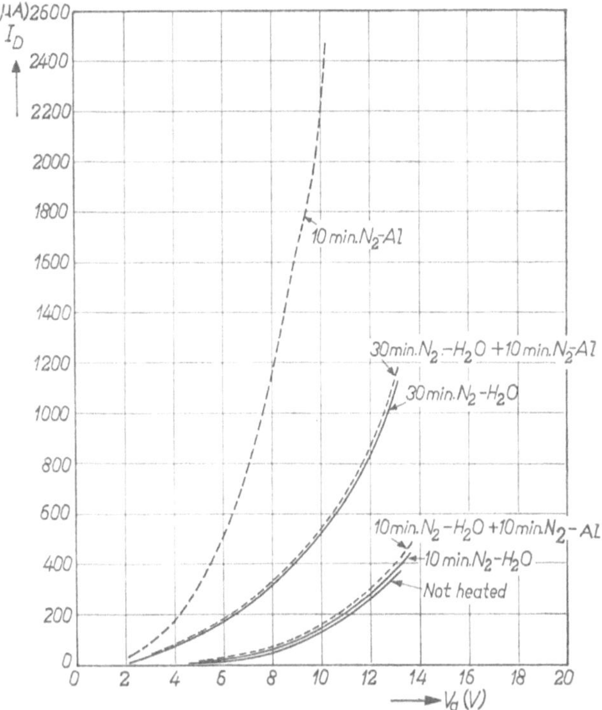

Fig. 4.4. Measurements on *npn*-type MOS transistors, with an oxide made in dry O_2 at 1200 °C. The drain-current versus gate voltage is shown before and after heat treatment in dry nitrogen at 300 °C for 10 minutes, with an Al electrode present. These treatments had no effect on the characteristics when a pre-heat treatment in wet N_2 at 300 °C had been given.

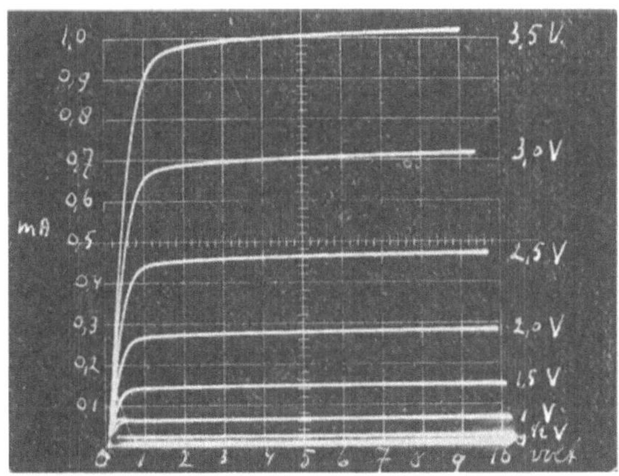

Fig. 4.5. I_D-V_G characteristics of an *npn* MOS transistor with varying gate voltage. This sample had an oxide thickness of 0·2 μ and a source-drain distance of 18 μ.

oxide), the channel conductance at $V_G = 0$ is very low, but starts to increase at low positive gate voltages. In fig. 4.5 the drain characteristics of such a transistor have been shown for various values of V_G. In fig. 4.6 the C-V curve, measured at a frequency of 500 kc/s, has also been given. This curve indicates that for zero band bending a considerable amount of negative charge (1 volt corresponds to about 10^{11} unit charges per cm^2) still has to be applied to the gate electrode.

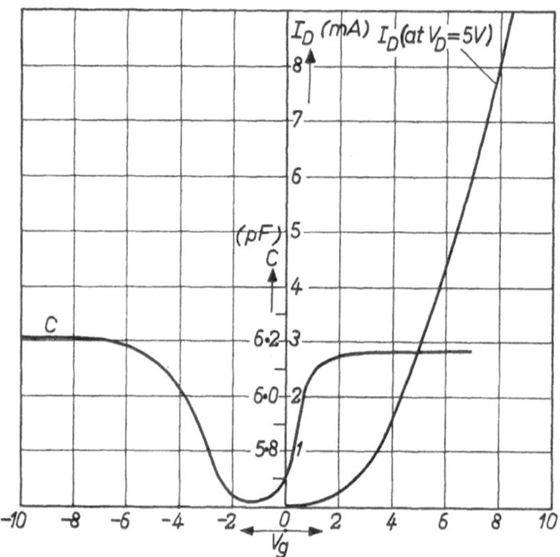

Fig. 4.6. High-frequency (500 kc/s) C-V_G curve and the I_D-V_G curve of the MOS transistor of fig. 4.5.

4.5.2. *pnp structures*

Measurements of the capacitance curve and the drain current as a function of gate voltage are presented in fig. 4.7. These measurements point to the presence of surface states, which must be considered as hole-trapping centres. Again the transconductance could be increased by heating in hydrogen or water vapour. In fig. 4.7 the C-V and I_D-V_G curves have also been given after a heat treatment in wet nitrogen at 450 °C for 15 minutes. The presence of an aluminium electrode during heating had a similar effect, in accordance with the experiments of Cheroff et al. [4-4]).

The large similarity in behaviour of the oxide-covered *npn* and *pnp* samples during heat treatment indicates that both the electron-trapping surface states and the hole-trapping surface states disappear due to a reaction of the centres to which they belong with hydrogen atoms. It may well be that the same centres can act in one case as a trap for electrons and in another case as a trap for holes, depending on the electron and hole concentrations at the surface of the silicon

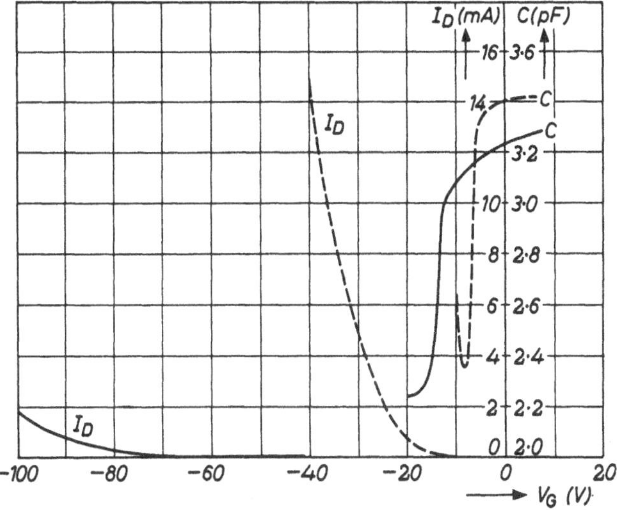

Fig. 4.7. High-frequency (500 kc/s) C-V_G curve and the I_D-V_G curve of a *pnp*-MOS-transistor structure, made before (drawn curve) and after (dashed curve) heat treatment of the oxidized sample in wet nitrogen at 450 °C for 15 minutes. The oxide had been made by oxidation in dry oxygen at 1200 °C during 1 hour. Source-drain distance: 25 microns.

crystal. It is interesting to note that due to the reaction with hydrogen the I_D-V_G curves showed much larger shifts than the accumulation-depletion region of the C-V_G curves (tables 4-I and 4-II and fig. 4.7). This indicates that the centres tend to become positive when large negative gate voltages are applied and tend to become negative when the gate voltage is made positive.

4.6. Surface charge, present after annihilation of the fast surface states

The C-V_G as well as the I_D-V_G curves measured after annihilation of fast surface states indicate that in the experiments described in this paper some positive charge always remained present at the surface. Although it is questionable whether all interface states were removed in these treatments, it is plausible to associate the remaining displacement of the C-V_G and I_D-V_G curves from the theoretical cases of no surface charge to a charge distribution between oxide and silicon.

The results, given for the oxides covered by the phosphate glass, appeared to be very reproducible. The lowest surface charge was obtained after heat treatment in water vapour (table 4-I) and was equal to 2.10^{11} unit charges per cm^2 with a spread from one experiment to another of less than 2.10^{10}. This reproducibility suggests that the positive charge is not due to the presence of accidentally introduced impurities but rather to some fundamental property of the oxide or its boundary with the silicon. The reproducibility was less favourable when the oxide had not been subjected to a phosphorus diffusion, which suggests that the phosphate glass has a gettering action for certain impurities. (In chapter 7 it will be shown that the presence of sodium impurities can have a considerable effect on the interface properties, and that such impurities can be gettered by the phosphate glass).

It can be argued (see chapter 7) that V_f, as indicated in table 4-I, gives a qualitative indication for the effective amount of oxide charge. One may assume that this positive charge is related to the oxygen-deficient state of the part of the oxide film present very close to the interface or, in other words, to the presence of donor-like unsaturated silicon bonds in the oxide structure. A reaction of these bonds with hydrogen may explain the decrease of V_f during various treatments mentioned in table 4-I. It may be noted that V_f increases due to treatment in oxygen. This may be explained by diffusion of traces of hydrogen, already present at the Si-SiO_2 interface, towards the oxygen-rich top layer of the oxide. This would cause an increase of the number of unsaturated silicon bonds in the oxide near the interface and thus result in an increase of the positive oxide charge.

Further considerations on the relationship between the structure of the oxide and certain properties of oxidized silicon will be given in the chapters 6 and 7. In chapter 6, which deals with effects of ionizing irradiations on oxidized samples, it will be shown that although the effect of water vapour and hydrogen on the interface states may be comparable, the effect on the oxide structure is different.

4.7. Conclusions

In oxidized silicon charge distributions in the oxide-silicon system have an important bearing on the properties of the surfaces. Apart from fixed charge in the oxide film, in many cases interface centres are present with which charge in the silicon can easily be exchanged. In this chapter we have deduced their presence by correlating C-V_G and I_D-V_G measurements of MOS-transistor structures. The experiments indicate that in general a larger number of these centres is present in samples which have been oxidized or heat treated in a dry ambient than in samples oxidized in a wet ambient. The number of the surface states is further increased when the oxidized samples are subjected to a phosphorus diffusion in a dry ambient, so that the top layer of the oxide film is converted to a phosphate glass.

In this chapter special consideration has been given to the latter types of samples. The phosphorus diffusion was carried out in nitrogen. A comparison of these samples with those made in a dry or wet ambient without phosphorus diffusion suggests that during the phosphorus diffusion the oxide becomes extremely dry. Both the large number of surface states, the low oxide conductance and the absence of hysteresis effects in C-V measurements [4-1]) of the phosphorus-containing samples indicate this. A part of these properties may also be caused by a gettering action of the phosphate glass for certain impurities.

The influence of hydrogen on the number of surface states is confirmed by the fact that during low-temperature heat treatments of oxide-covered samples in hydrogen the number of surface states decreases considerably. Water vapour has a similar effect, but in this case the effect is slowed down by the presence of oxygen, whereas it is enhanced when the oxide is in a reduced state. It is suggested that an aluminium electrode, present on top of the oxide during heat treatment, can be considered as a reducing agent for the oxide or for water present at its surface. Consequently hydrogen atoms may form at the aluminium-oxide interface, when one does not work under extremely dry conditions, and diffuse through the oxide and react with the trapping centres, supposed to consist of unsaturated silicon bonds, near the oxide-silicon interface.

The effect of water vapour and of an aluminium electrode on top of the oxide film during heat treatment can be slowed down by making centres in the oxide film in which hydrogen atoms can be trapped instead of reaching the silicon surface. The presence of a phosphate glass on top of an oxide has such an action. Also a pre-heat treatment in oxygen and under certain conditions water vapour at a low temperature can have such an effect.

Surface states are found on *p*-type as well as on *n*-type samples after oxidation in a dry ambient. In *npn* MOS transistors they influence the dependence of the channel current on the gate voltage because they are able to trap electrons. In *pnp* samples the presence of hole-trapping centres has an influence on the transistor characteristics. After annihilation of the interface states a fixed positive surface charge always appears to be present. The exact nature of the various centres which determine the surface properties of oxidized silicon has not yet been found. It seems probable that the active surface states are related to unsaturated silicon bonds near the oxide-silicon interface, which can loose their electron- or hole-trapping action after they have become saturated with hydrogen. A more extensive model of various possible centres at the Si-SiO_2 interface will be given in chapter 7.

REFERENCES

[4-1]) M. V. Whelan, Philips Res. Repts **20**, 562-577, 1965.
[4-2]) J. E. Thomas Jr and D. R. Young, IBM J. Res. Dev. **8**, 368-375, 1964.
[4-3]) D. P. Seraphim, A. E. Brennemann, F. M. d'Heurle and H. L. Friedman, IBM J. Res. Dev. **8**, 400-409, 1964.
[4-4]) G. Cheroff, F. Fang and F. Hochberg, IBM J. Res. Dev. **8**, 416-421, 1964.
[4-5]) H. S. Lehman, IBM J. Res. Dev. **8**, 422-426, 1964.
[4-6]) K. H. Zaininger and G. Warfield, Proc. IEEE **52**, 972-973, 1964.
[4-7]) J. Olmstead, J. Scott and P. Kuznetzoff, IEEE Trans. **ED-12**, 105-107, 1965.
[4-8]) A. B. Kuper and E. H. Nicollian, J. electrochem. Soc. **112**, 528-530, 1965.
[4-9]) P. Balk, Paper given at the Spring Meeting of the Electrochemical Society, San Francisco, 1965, Electronics Division, Abstract 109.
[4-10]) E. Kooi, J. electrochem. Soc. **111**, 1383-1387, 1964 (chapter 3 of this book).
[4-11]) D. R. Kerr, J. S. Logan, P. J. Burkhardt and W. A. Pliskin, IBM J. Res. Dev. **8**, 376-384, 1964.
[4-12]) A. G. Revesz, IEEE Trans. **ED-12**, 97-101, 1965.

5. INFLUENCE OF X-RAY IRRADIATIONS ON THE CHARGE DISTRIBUTIONS IN METAL-OXIDE-SILICON STRUCTURES *)

Abstract

Insulated-gate silicon field-effect transistors were irradiated by X-rays. The properties of the devices altered considerably depending on the total exposure to irradiation and the gate voltage as applied during the irradiation. The alterations can be explained by assuming that electric currents can flow in the MOS system under these conditions. This causes charge redistributions, which can be locked in at the moment that the irradiation is stopped. Due to these effects MOS devices may offer certain possibilities for detection of ionizing radiations.

5.1. Introduction

In the investigations to be described use has been made of MOS (metal-oxide-semiconductor) transistors of circular structure. A cross-section is shown in fig. 5.1. They were made by diffusion of phosphorus into a *p*-type silicon substrate, using oxide-masking techniques. In such a structure a conducting *n*-type channel is often present between the "source" and "drain" regions, even if no voltage is applied to the metal gate contact. This is due to a charge distribution between the silicon and centres in the oxide or at the silicon-oxide interface. When the gate contact is made positive with respect to the silicon, an additive negative charge is built up in the silicon and the channel conductance is thus increased. The increase also depends on the mobility of electrons in the *n*-type channel and the number of surface states in which the electrons may become trapped. The channel conductance can be decreased by applying a negative potential to the gate electrode.

The amount of charge induced at a silicon surface by the application of a certain gate voltage can alter in the course of time if charge carriers are able to move in the oxide film, or if charge transfer takes place between centres in the oxide and silicon. In many cases these effects are slow at room temperature,

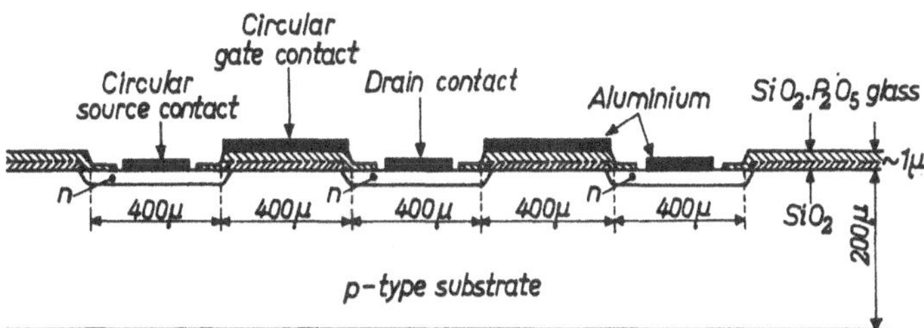

Fig. 5.1. A cross-section of the circular MOS-transistor structure used in the experiments.

*) Published: Philips Res. Repts **20**, 306-314, 1965.

but can have a considerable influence at somewhat higher temperatures [5-1,2]). Apart from the nature of the charge carriers in the oxide film, the charge-transfer processes at the metal-oxide and the oxide-silicon boundary can also have a marked influence on the change in the charge distributions [5-3]), which can be locked in during cooling to room temperature.

We have observed that the charge distributions at oxidized silicon surfaces may also be changed considerably if they are subjected to ionizing irradiations. In this paper we shall discuss the influence which the presence of electric fields across the oxide can have during X-ray irradiations. The experiments were done at room temperature, so that any change in the characteristics of the field-effect transistors could be observed immediately. We think that most of the observed changes can be explained by assuming that, as a result of ionization processes, electrons are able to move in the oxide. The results indicate that in these irradiation experiments, too, the bottleneck of charge transfer in the MOS structure often lies at some boundary in the system.

5.2. Experimental procedure

The MOS transistors, whose dimensions are given in fig. 5.1, were made by oxide-masking techniques. The starting material was 5-Ωcm indium-doped floating-zone *p*-type silicon. Slices were cut perpendicular to the ⟨111⟩ axis of the crystal. After lapping with fine alundum powder, the wafers were etched in an HF-HNO_3 mixture to remove a layer of about 60 microns from each side. Thermal oxidation was carried out at 1200 °C for 16 hours in wet oxygen, made by bubbling oxygen through water at 30 °C. Photo-engraving techniques were used to make windows in the oxide film with a view to producing *n*-type source and drain regions by phosphorus diffusion, which was carried out in nitrogen. Phosphorus was deposited for 30 minutes in a two-zone open-tube diffusion system with a P_2O_5 source held at 220 °C and silicon at 920 °C. The silicon slices were then heated at 1150 °C for 4 hours. In this time *n*-type diffused regions were formed with a *pn*-junction depth of 6 microns. The samples were cooled rather slowly by turning off the power supply to the furnace. The total oxide thickness obtained in this process was about 1·2 microns.

In this stage the samples showed little channel conductance, probably owing to electron-trapping effects at the surface. Such effects can often be diminished by suitable annealing treatments (see chapter 4). We applied treatments in air or oxygen at 500 °C to make samples with various channel conductances. Source, drain and gate contacts were then made by vapour deposition of aluminium and a subsequent photo-engraving process. The *p*-type substrate was alloyed to a gold-plated transistor header, and gold wires were attached to the source, drain and gate electrodes. The devices were capsulated in dry air.

The X-rays were obtained from an X-ray tube with a tungsten anode, to which

150 kV was applied. The radiation was sufficiently energetic to penetrate through the metal can of the transistor header. The exposure rate of the irradiations applied to the transistors was 10^4 röntgens per minute. During the irradiations the voltage between the gate and the source electrode was variable. Source, drain and *p*-type substrate were externally connected, except during short intermediate measurements, during which a voltage was applied between source and drain contacts to be able to observe alterations in the channel conductance. If such a voltage had been maintained during the irradiations, the results would have been different, because the field distribution at the surface would have been affected.

5.3. Results and discussion

Figure 5.2 shows the drain characteristics of a non-irradiated transistor. The drain current I_D has been plotted against the drain voltage V_D between drain and source, while the gate voltage V_G between gate and source is a separate parameter. The source and the *p*-type substrate were short-circuited. Before metallization this sample had been subjected to a heat treatment in air at 500 °C for 15 minutes.

Figure 5.3 gives an example of what happened when an X-ray treatment was given to such a sample with a constant gate voltage applied, in this case equal to —2 V. The drain current (at a drain voltage of 25 V) was measured from time to time in the course of the irradiation. Considerable changes were observed, until after a few minutes a more or less stable value had been reached. The channel conductance was seen to alter again when the gate voltage was changed, until a new saturation value had been reached. The saturation values of the drain currents were measured for various values of the gate voltage applied during the irradiations. The results are plotted in fig. 5.4 (points along the dashed line). When positive gate voltages were applied during the irradiations, the channel conductance became so large that no saturation value of the drain current (at $V_D = 25$ V) could be measured.

Another remarkable effect was that the changes in charge distribution in the MOS system, to which the alterations in the drain currents are correlated, could be locked in at the moment that the irradiations were stopped. That is to say, the changes then became markedly smaller. This is demonstrated in fig. 5.3 for a case where the saturation value of the drain current for the given gate voltage had already been reached during irradiation.

The marked storage effects which the devices showed for the conditions of irradiation allowed us to measure the drain characteristics of the transistor for various induced charge distributions in the MOS system. Results have been plotted in fig. 5.4. The dotted line shows the dependence of the drain current (at $V_D = 25$ V) on the gate voltage before the sample was irradiated. The drawn

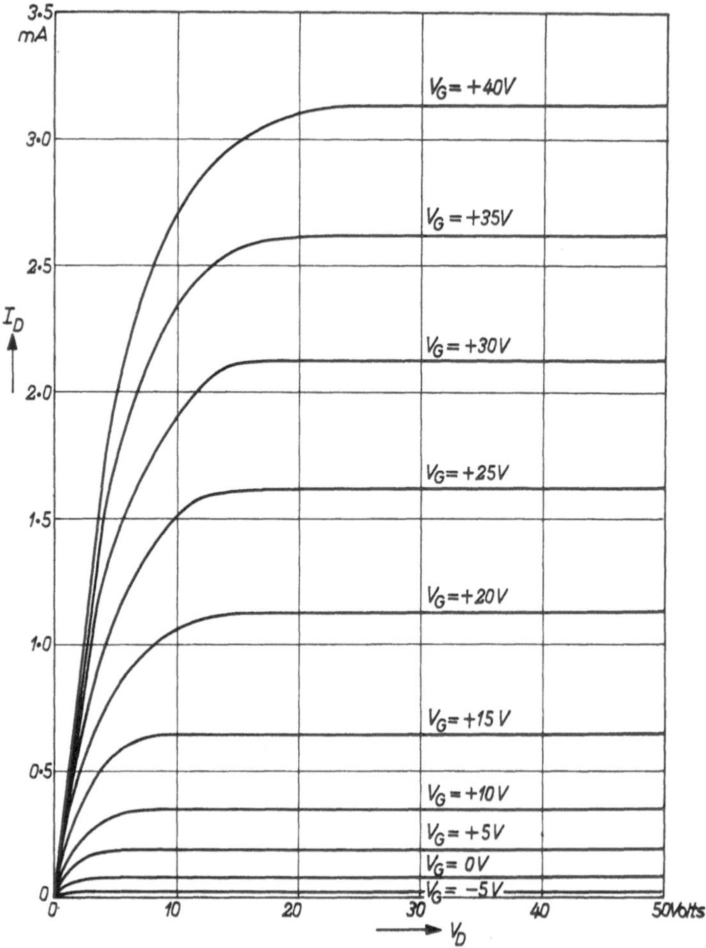

Fig. 5.2. Drain characteristics of a MOS transistor;
V_D = voltage on the drain contact with respect to the source,
I_D = current through the drain contact,
V_G = voltage on the gate electrode with respect to the source.

curves were measured after various irradiations. In these cases the irradiations had been stopped after a more or less stable charge distribution at a given gate voltage had been reached. The values of these gate voltages are indicated on the various lines. The parallelism of these lines indicates that the transductance (defined as dI_D/dV_G) was not markedly influenced by the irradiation process, except perhaps at low drain-current values. The effects of the irradiations may therefore indeed be ascribed to changes in the amount of locked surface charge or its distribution in the oxide film. Neglecting interface states we may say that these changes are directly correlated to the values of the drain current at a certain gate voltage, for example at $V_G = 0$. Figure 5.4 shows that the drain

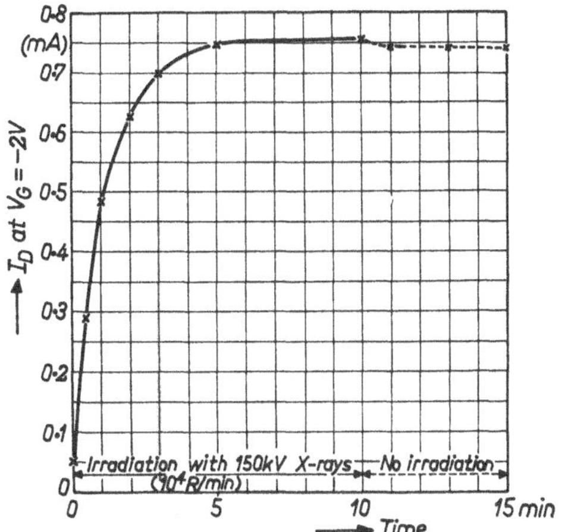

Fig. 5.3. A typical example of the dependence of the drain current on the time of irradiation with the gate voltage kept constant (in this case at —2 V). The figure shows also the behaviour of the drain current during 5 minutes after the irradiation had been stopped.

current at $V_G = 0$ exhibited an increase both when positive and large negative gate voltages had been applied during irradiation. In this device a minimum in the drain current at $V_G = 0$ was obtained when the gate voltage during irradiation was maintained at —20 V. Note that the irradiations never resulted in a *p*-type surface (at $V_G = 0$).

In fig. 5.5 the same kind of measurements has been plotted for another device. This device was one from a series that had been made in the same way, except for the low-temperature annealing being carried out before metallization, which had now been done in oxygen at 500 °C for 60 minutes. The device already showed a fairly high channel conduction at $V_G = 0$ before irradiation (dotted curve in fig. 5.5). Its behaviour during irradiation is comparable to that of the previous device. Nevertheless fig. 5.5 shows a remarkable difference compared with fig. 5.4. In fig. 5.4 the saturation value of the drain current was equal to zero when large negative gate voltages were applied during the irradiation. In fig. 5.5, however, the saturation value remains practically constant at a rather high value of the drain current over a wide range of applied gate voltages during irradiation.

The effects are most readily explained by considering that a voltage applied across a MOS structure can cause charge redistributions only if electric currents can flow. In a stationary state the net electric current will be the same everywhere in the circuit. Therefore the largest electric field tends to arise in that part of the system where charge transport is most difficult. The measurements done during the irradiations indicate that a stationary value of the drain current could often be reached for a given applied gate voltage. The drain current is correlated to the surface potential of the silicon, and thus one may also say that in the

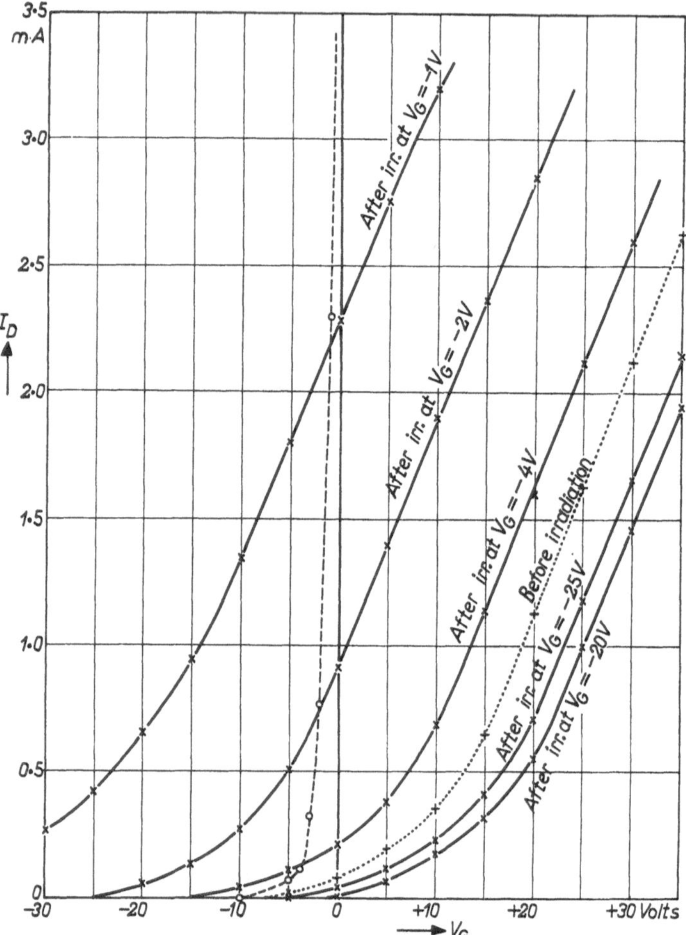

Fig. 5.4. The drain current of a MOS transistor as a function of the gate voltage, while the drain voltage was held constant (25 V); dotted curve: before irradiation; dashed curve: saturation values obtained during irradiation by X-rays at an exposure rate of 10^4 R/min, measured at the same gate voltage as applied during irradiation; drawn curves: after irradiation with the indicated gate voltage applied until a stationary state at this voltage had been reached.
In the procedure for making this sample, a heat treatment during 15 minutes at 500 °C in air was an essential part.

stationary states obtained during various irradiation experiments the electric field at the silicon surface reached a saturation value depending on the irradiation conditions.

When positive gate voltages were applied during the irradiations, the drain currents, and thus the electric fields at the O-S boundary, became very large and were highly sensitive to the applied voltage. This indicates that charge transfer from silicon to the oxide was difficult in these cases. It may be assumed that

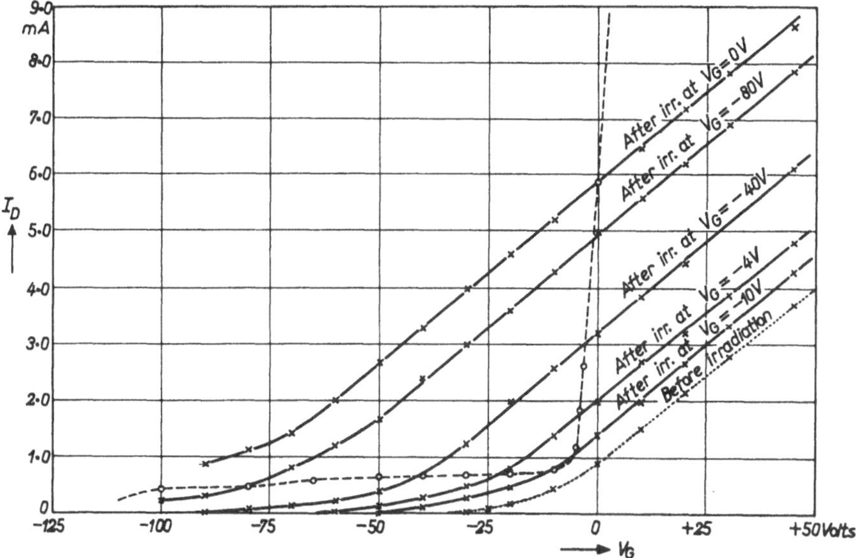

Fig. 5.5. The drain current of a MOS transistor as a function of the gate voltage, while the drain voltage was held constant (50 V); dotted curve: before irradiation; dashed curve: saturation values obtained during irradiation by X-rays at an exposure rate of 10^4 R/min, measured at the same gate voltage as applied during irradiation; drawn curves: after irradiation until a stationary state at the indicated gate voltage had been reached.
The samples differ from that of fig. 5.4 in the low-temperature heat treatment at the end of the diffusion process. In this case annealing had been carried out during 1 hour in oxygen at 500 °C.

X-rays induce free electrons in the oxide which can move to the positive metal contact and possibly be carried off there. Presumably the X-rays also create in the silicon a number of electrons at a sufficiently high energy level to enter the oxide film *), but apparently this photocurrent is not sufficient to compensate the electron flow in the oxide in the direction of the gate contact. Thus, a positive space charge can be formed in the oxide film, while electrons accumulate in the inversion layer in the silicon. In this way the field at the O-S barrier increases, whereas the field in the remaining part of the oxide film decreases until a stationary value of the electric current throughout the MOS system has been reached.

When the gate voltages were increased in the negative direction, the irradiations could also induce an increasing amount of positive charge in the oxide. This remarkable effect may be explained by assuming that in these cases elec-

*) In other experiments with oxidized silicon surfaces we have noted that electron transfer from silicon to oxide can also be stimulated by illumination with u.v. light with an energy of more than 4·2 eV (see chapter 6).

trons can flow rather easily from the oxide to the silicon, while this flow cannot be compensated by the photocurrent at the M-O barrier until in a stationary state the charge distribution has changed in such a way that the current is again the same everywhere in the system. If electron transport at the M-O boundary is indeed very difficult, the electric current across this barrier possibly remains of the same order of magnitude when the electric field increases. An increase of the gate voltage in the negative direction may then (after sufficiently prolonged irradiation) have only little effect on the electric field at the O-S boundary, because finally about the same current has to flow. This would explain the nearly horizontal part of the dashed curve in fig. 5.5, which indicates that the electric field at the silicon surface reached finally about the same value during irradiation, whereas the total voltage varied over a wide range.

A comparison of figs 5.4 and 5.5 shows that the process by which the samples were prepared can influence the properties of the devices both before and during irradiation. Both figures show that for small negative voltages the effects of irradiation were such that it seemed as if (in terms of the given explanations) a positive voltage was applied. At zero voltage there is probably still a net electron current from the oxide to the metal electrode. A certain negative voltage has to be applied to stop this current. When the negative voltage is increased further the current becomes determined by the amount of electrons excited by the X-rays in the metal to a sufficient high energy to enter the oxide. The amount of these electrons will not be affected by the electric field at the M-O boundary and so the current becomes nearly independent of the gate voltage. At which value of the negative gate voltage this will occur depends on the nature of the metal and the properties of the oxide near the metal. As such the presence of a phosphate glass on top of the oxide may also be of importance.

The way of preparation of the oxide film also appeared to have a marked influence on the behaviour of oxide-protected silicon devices which were irradiated during longer periods than those applied here. We have observed that in many cases the saturation values of the drain currents obtained at a given gate voltage altered again during continued irradiations. We think that this is due to the formation of new centres in the oxide or at the oxide-silicon interface. Presence of such surface states affects in many cases also the dependence of the drain current on the gate voltage. Negative charge induced at the silicon surface may become trapped in surface states, instead of giving a contribution to the channel conductance. Consequently continued irradiation of MOS transistors may influence their transconductance (given by the slope of the drawn curves in figs 5.4 and 5.5). Such phenomena have been observed previously by irradiation of oxidized silicon surfaces with γ-rays [5-4,5]). We have found that these effects can be very pronounced if the oxide films have been made or heat treated in a wet ambient (see next chapter).

5.4. Possible application for detection of ionizing irradiation

The experimental results suggest that MOS devices may possibly have certain merits for use in dosimetry. Apart from X-rays, other kinds of irradiation, such as ultra-violet light, could also possibly be detected, provided provisions are made to ensure that the radiation can enter the oxide below the gate electrode. The sensitivity of the devices can be influenced by varying the gate voltage, although it depends also on the stability of the drain current when no irradiation is applied. With our devices we were still able to see noticeable influence with exposure rates of X-rays as low as 0·01 R/min.

5.5. Conclusions

When oxidized silicon surfaces are subjected to ionizing irradiations, a redistribution of charge may occur. The effects depend on the way in which the oxide films have been prepared and also on the presence of electric fields such as can easily be applied in a MOS structure. In the devices which we used, an increasing effect of positive surface charge was found by the application of either a positive or a large negative potential on the metal electrode during irradiation by X-rays. The charge can become locked in the oxide film after the irradiation has ceased. During continued irradiation the sensitivity of the device for radiation may change, due to structural changes in the oxide-silicon system.

REFERENCES

[5-1]) M. Yamin and F. L. Worthing, Conference of the electrochemical Society, Toronto, 1964; Extended Abstracts of the Electronics Division **13**, 182, 1964.

[5-2]) J. E. Thomas Jr and D. R. Young, I.B.M. J. Res. Dev. **8**, 368-375, 1964.

[5-3]) D. P. Seraphim, A. E. Brenneman, F. M. d'Heurle and H. L. Friedman, IBM J. Res. Dev. **8**, 400-409, 1964.

[5-4]) H. L. Hughes and R. G. Giroux, Electronics **37**, 58-60, 1964.

[5-5]) H. Edagawa, Y. Morita, H. Ishikawa, S. Maekawa and Y. Inuishi, Jap. J. appl. Phys. **3**, 644-660, 1964.

6. EFFECTS OF IONIZING IRRADIATIONS ON THE PROPERTIES OF OXIDE-COVERED SILICON SURFACES *)

Abstract

The influence of ionizing irradiations on the surface properties of oxidized silicon was measured by observing changes in the conductance of inversion layers, present at the surface of *p*-type silicon. In a number of cases high-frequency capacitance versus voltage measurements of MOS structures were employed too. From both kinds of measurements it is concluded that in addition to an effect on the charge distribution in the oxide-silicon system also new centres may form at the surface during irradiation. The latter effect was found to be most pronounced in samples which had been prepared in a wet ambient. In these cases the *n*-type character of the surface often tended to disappear during continued irradiation. Certain possible relations between the structure of SiO_2 and the irradiation effects are discussed.

6.1. Introduction

In the previous chapter it has been shown that X-ray irradiations on MOS (metal-oxide-semiconductor) structures may result in charge redistributions which depend on the voltage across the system. In those experiments it was also noted that the number of centres in the oxide or at the oxide-silicon interface altered in the course of an irradiation. This is in accordance with the decrease in transconductance of MOS transistors, due to γ-ray irradiations, which was observed by Hughes and Giroux [6-2]. Edagawa et al. [6-3] found that *n*-type inversion layers present at the surface of oxidized *p*-type silicon disappeared during γ-ray irradiations. They also observed changes in the infrared absorption of the oxide films [6-4], which pointed to structural changes. These films had been made by thermal oxidation in steam. In this chapter it will be shown that irradiation by X-rays or ultra-violet light can also cause the disappearance of inversion layers on *p*-type silicon present after oxidation in a wet ambient. The way of oxide preparation will be shown to have a marked influence on the effect of ionizing irradiations. Certain possible relations, correlations between the structure of the oxide films, the surface properties of the underlying silicon and the radiation effects will be discussed in sec. 6.4.

6.2. Experimental procedure

In general the sample preparation was similar to that given in chapter 4. In many cases, however, the oxide was removed after the phosphorus-diffusion process and a new oxide was then grown in either a wet or a dry ambient. The structure which was then obtained is shown in fig. 6.1. This is equivalent to the MOS-transistor structure used in the work discussed in the preceding chapters,

---

*) Published: Philips Res. Repts **20**, 595-619, 1965.

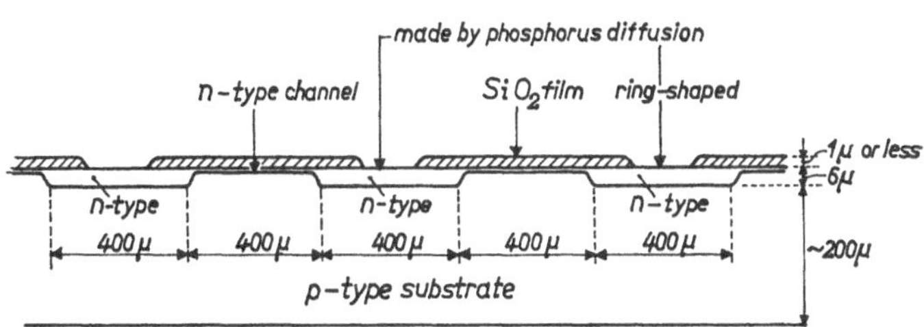

Fig. 6.1. Cross-section of a sample used for measurements of channel currents through inversion layers at the surface of oxidized *p*-type silicon.

but without the metal gate contact. During the irradiations such a metal layer was generally not present either, but in a few cases it was applied after the irradiation was stopped in order to be able to measure the channel conductance or high-frequency (500 kc/s) capacitance versus voltage characteristics of MOS structures. In other cases only the channel currents were measured, without applying a gate electrode.

The X-rays were obtained from an X-ray tube with a tungsten anode. Results will be given for experiments with an anode voltage of 150 kV (a lower voltage resulted in similar although somewhat smaller effects). During the irradiations the oxidized surfaces were directed to the X-ray source. A few experiments were done with γ-rays from a Co^{60} source. Ultra-violet illuminations were carried out by placing the samples close to a 15-W low-pressure-mercury spectral lamp (Philips 93109). The dominant light emission of this lamp lies at 2537 Å. Near the samples the intensity of the light at this wavelength was in the order of 10^{19} photons per second per cm^2. We have also employed illuminations from a 1600-W xenon lamp followed by a monochromator with quartz prisms. In these cases we intended to study the spectral response of the light effect. As the intensity was rather small under these conditions (about 10^{13} photons per second per cm^2 at 2537 Å), we could use this method only in a few cases.

6.3. Results of irradiation experiments

6.3.1. "*Wet*" *and* "*dry*" *oxides*

6.3.1.1. Irradiation by X-rays

In fig. 6.2 and 6.3 the influence of X-ray irradiations on channel leakage currents has been illustrated for a number of oxide-covered *npn* structures as shown in fig. 6.1. The samples had been reoxidized after removal of the oxide at the end of the phosphorus diffusion in an aqueous HF solution. The behaviour of a number of samples covered by an oxide, made in a wet ambient, is shown in fig. 6.2. During the first period of irradiation the leakage currents increased, but soon a maximum was reached, and then the channel conductance decreased

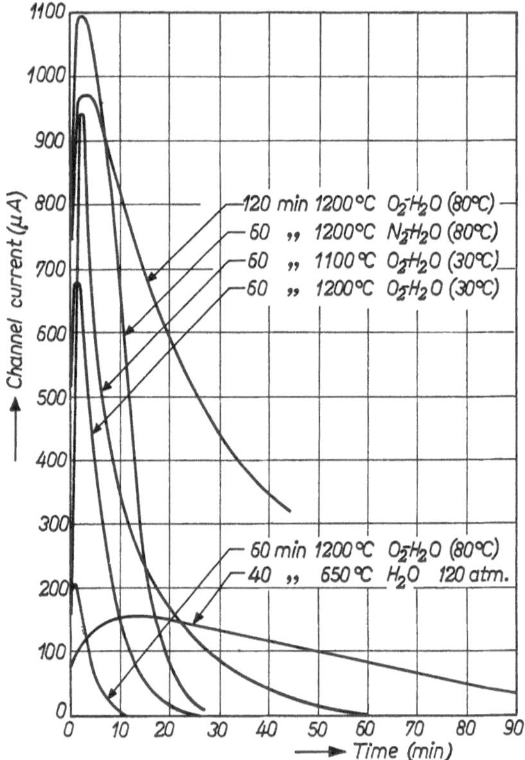

Fig. 6.2. Influence of X-ray irradiations (150 kV, exposure rate 10^4 R/min) on the channel currents through inversion layers at the surface of 5-Ωcm *p*-type silicon, oxidized in various wet ambients. The oxidation conditions are given along the various curves, which were found by measurements after various periods of irradiations.

again, after some time leading to a channel current which was much lower than the starting value.

The samples which had been reoxidized in dry oxygen showed much smaller channel currents than those covered by "wet" oxides. During the first period of irradiation the channel currents increased but after a few minutes an equilibrium value was reached and further change was very small. A few examples have been given in fig. 6.3. "Wet" oxides, which had been reheated in a dry ambient (O_2 or N_2) at the oxidation temperature behaved as "dry" oxides during X-ray irradiation.

In a few cases we also measured the capacitance versus voltage characteristics of MOS structures made before and after irradiation of oxidized samples by vapour deposition of aluminium spots (diameter 1 mm) on the oxide film. Figure 6.4 shows the influence of X-ray irradiation on the *C-V* curve of a sample oxidized in a wet ambient; fig. 6.5 shows the same effect for a sample oxidized in dry O_2. In both cases a shift of the *C-V* curve to a more negative voltage can

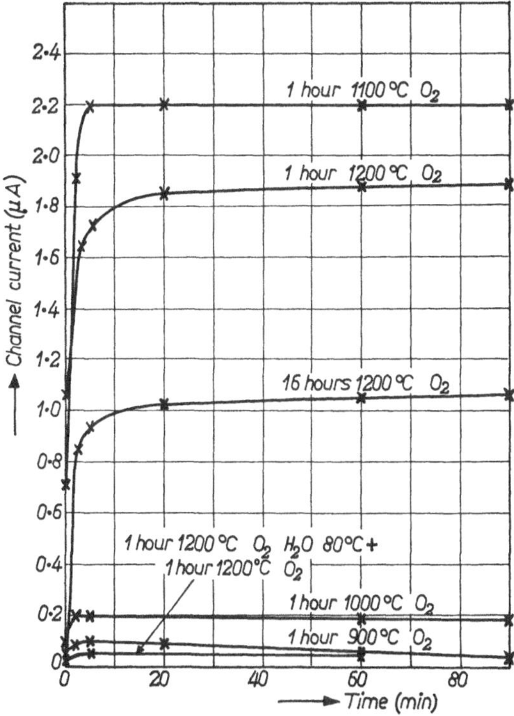

Fig. 6.3. Influence of X-ray irradiations (exposure rate 10^4 R/min) on the channel currents through inversion layers at the surface of 5-Ωcm p-type silicon, oxidized in dry oxygen.

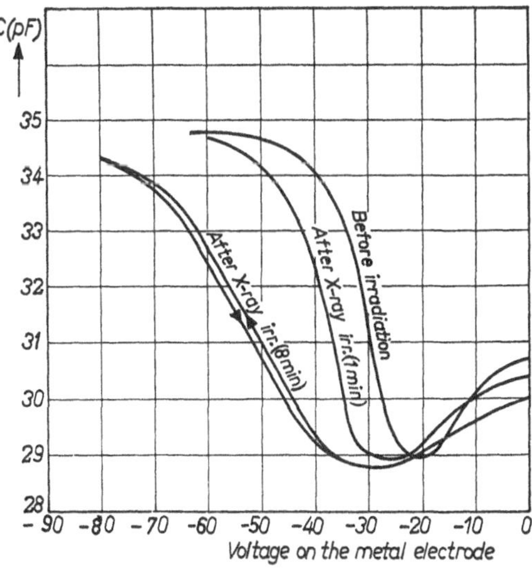

Fig. 6.4. Measurements of the C-V curve (at 500 kc/s) of MOS structures (diameter 1 mm) before and after irradiation of a 5-Ωcm p-type silicon slice, which has been oxidized for 120 min at 1200 °C in oxygen, saturated with water vapour at 80 °C. The hysteresis effect shown in the curve on the left-hand side was found by measuring first from positive to negative gate voltages and reverse after maintaining —120 V on the metal contact for 3 minutes.

be noticed. For the sample with the "wet" oxide there is a secondary effect: the slope of the *C-V* curve becomes markedly smaller after a certain period of the irradiation. The relation between the changes of the *C-V* characteristics of figs 6.4 and 6.5 and the channel currents of figs 6.2 and 6.3 will be discussed in

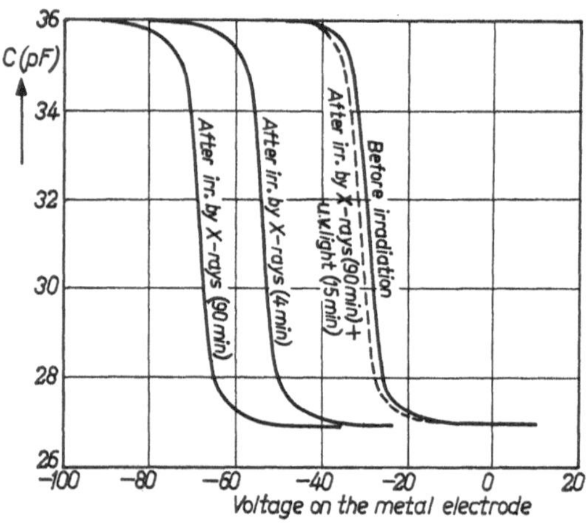

Fig. 6.5. Shifts of the *C-V* curves (measured at 500 kc/s) of MOS structures (diameter 1 mm) as a consequence of X-ray irradiation on samples oxidized in dry oxygen for 16 hours at 1200 °C. The shift-back due to illumination by u.v. light after an X-ray irradiation is also shown.

sec. 6.4. We have also noted that a hysteresis effect in the *C-V* curves was induced during irradiations, in particular for samples covered by a "wet" oxide. In terms of Whelan's interpretations [6-5]) the type of hysteresis as shown in fig. 6.4 is due to slow charge transport between centres in the oxide and silicon or vice versa.

In this section general rules were given for the irradiation effects on "wet" and "dry" oxides. It is felt (and will be discussed more thoroughly in sec. 6.4) that the primary effect of irradiation is a charge redistribution between oxide and silicon, while secondary effects are caused by formation of new defects.

6.3.1.2. Irradiation by γ-rays

The effect of γ-ray irradiations was studied in a number of cases. An exposure to 2.10^7 R resulted in a large decrease of the leakage currents when the oxidation had been carried out in a wet ambient. For example a channel current of 100 μA obtained during reoxidation in a wet ambient at 1200 °C became 1μA after the irradiation. The irradiation effects on the channel currents of samples with "dry" oxides were of the same order of magnitude as obtained during X-ray irradiations.

6.3.1.3. Irradiation by u.v. light

Figure 6.6 illustrates that the leakage currents through the *npn* structures when covered by "wet" oxides decreased considerably during illumination with u.v. light. However, little or no effect was found when the oxides had been grown in dry oxygen. The high-frequency *C-V* curves of MOS structures made before and after irradiations of samples with a "wet" oxide by deposition of aluminium

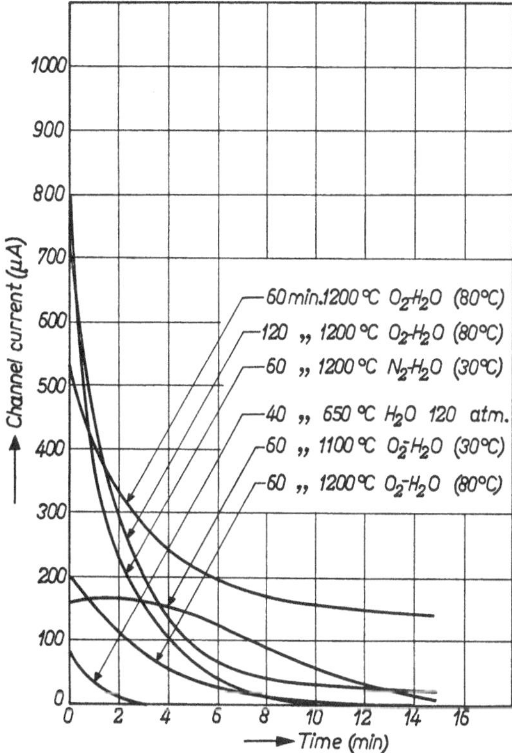

Fig. 6.6. Influence of u.v. light from a low-pressure-mercury lamp on the inversion layers present at the surface of *p*-type silicon samples, oxidized in various wet ambients.

spots, showed often (not always) a shift to less negative voltages, while also the slope decreased somewhat. In the case of "dry" oxides the influence of u.v. light on the *C-V* curves was negligible. The results indicate that whether or not u.v. light is able to cause a charge redistribution in the oxide-silicon system, depends strongly on the way in which the oxide is made.

6.3.1.4. Alternating X-ray and u.v. treatments

We have observed that the effect of u.v. light on the disappearance of *n*-type channels became often much larger when primarily an X-ray or γ-ray irradiation had been given. This effect was again very pronounced for samples covered by a "wet" oxide. An example has been given in fig. 6.7 for an *npn* structure covered by an oxide made by heating in wet oxygen (H_2O at 30 °C) for 10 minutes at 1100 °C. For comparison the effect of u.v. light on a non-X-rayed sample has also been shown.

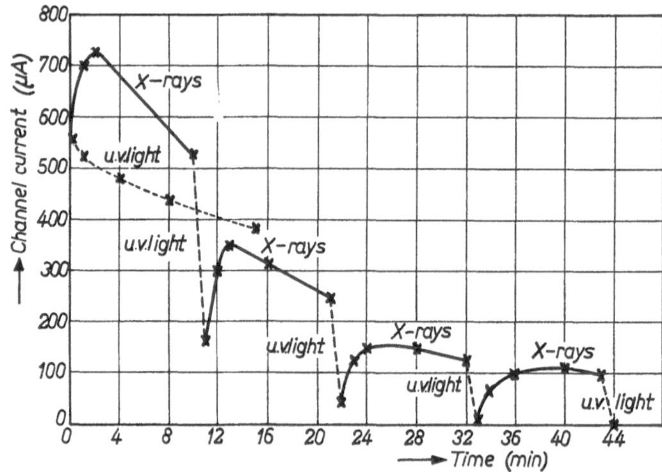

Fig. 6.7. Measurements of channel currents on a sample oxidized for 10 minutes in wet oxygen (water at 30 °C) after various irradiations by X-rays (150 kV-10^4 R/min) and u.v. light (from a low-pressure-mercury lamp).

The increase in channel currents during X-ray irradiation of samples which had been oxidized in dry oxygen disappeared again during a brief u.v. illumination. In the mean time *C-V* curves which were obtained during X-ray irradiation shifted back to about their original values. This effect has been illustrated in fig. 6.5.

6.3.2. *Oxide films covered by a phosphate glass*

In these cases oxidized surfaces were considered, which had been subjected to a phosphorus diffusion. The oxidation and diffusion procedures to make these films have been given in sec. 4.2.1. The final diffusion stage was a heat treatment for 4 hours at 1150 °C after which the samples were cooled rather slowly to room temperature. Channel currents are very low in such samples due to the presence of many electron-trapping centres at the surface, but can be increased by various heat treatments as was discussed in chapter 4. It will be shown that the effect of various irradiations depend on the nature of these treatments as well as on the presence of the phosphate glass upon the SiO_2 film.

6.3.2.1. Irradiation by X-rays

Channel currents of samples which had not been subjected to further heat treatments after the phosphorus diffusions, were initially only little influenced by

X-rays, but after several minutes of irradiation the n-type character increased. The reason for the slow increase during the starting period of irradiation becomes clear after observing the change in MOS-transistor characteristics of such samples, measured after applying a metal gate electrode before and after irradiation (fig. 6.8). In the same figure one can see how the high-frequency C-V_G

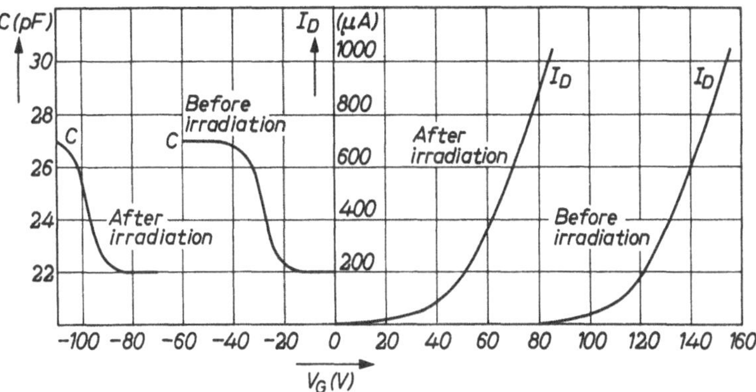

Fig. 6.8. I_D-V_G and C-V_G curves of a MOS transistor before and after exposure to X-rays (150 kV-10^4 R/min) for 15 minutes. During the irradiation no metal gate electrode was present.

curve of the MOS structure shifted during the same irradiation experiments. The conclusion is that, although the shift of both the I_D-V_G and C-V_G curves is considerable, the channel current (at $V_G = 0$) increases only to a small amount due to the presence of a large number of active trapping centres, which makes the slope of the I_D-V_G curve very small over a wide range of the gate voltage V_G.

Samples which had been heat treated in such a way (discussed in chapter 4) that the trapping centres at the silicon surface had disappeared, showed immediately a considerable increase of channel current during X-ray irradiation. This can be expected because the influence of the changes in surface charge on the channel currents is not masked now by the presence of the active surface states. We observed in these cases that during continued irradiation the slope of the I_D-V_G curve and the C-V_G curve often became smaller, indicating the formation of new surface states.

Frequently the results of irradiation became somewhat different when the phosphate glass had been removed from the samples. This is illustrated in fig. 6.9. The effect was less in the sample on which the phosphate glass was not present. Removal of the phosphate glass from the other wafer after one hour of irradiation did not have much effect on the channel current, but continued irradiation then caused a decrease of the channel current, until the two samples showed again about the same properties.

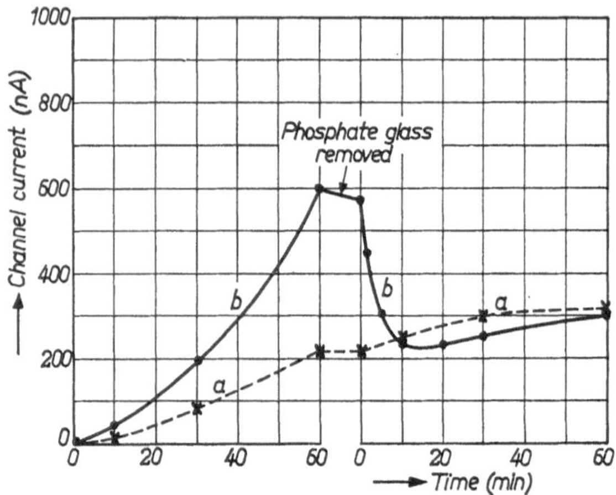

Fig. 6.9. Effect of the presence of a phosphate glass during irradiation by X-rays (150 kV-10^4 R/min). From sample a the phosphate glass had been removed before exposure, from sample b it was removed after 60 minutes of irradiation.

6.3.2.2. Irradiation by γ-rays

In these cases too, the surface covered by SiO_2 + phosphate glass always became stronger *n*-type. Together with other results a few examples can be found in table 6-II. The *C-V* curves of MOS structures were measured in the frequency range 50 kc/s to 5 Mc/s. They showed a shift to negative voltages but also their slope became considerably smaller, especially for the high-frequency curves, indicating formation of surface states with corresponding relaxation times. However, also hysteresis effects were found after irradiation, indicating the formation of much slower surface states.

6.3.2.3. Irradiation by u.v. light

The effect of u.v. light appeared to depend very much on the way in which low-temperature heat treatments after the phosphorus diffusion had been carried out. The effects of various heat treatments have been given in chapter 4. It was shown there that heat treatment at 450 °C in hydrogen or wet nitrogen, and also in dry nitrogen when an aluminium electrode was present on the oxide, resulted in the formation of *n*-type channels on the surface of the *p*-type material. This was explained as being due to the disappearance of electron-trapping centres rather than to a charge redistribution in the oxide-silicon system although this effect could not always be neglected either.

In fig. 6.10 a few results of u.v. illuminations on various samples have been given. Remarkable differences can be noted. The channel currents of the samples which had not been heat treated after the diffusion were not affected. The channels formed due to heat treatment in water vapour disappeared in the course of an irradiation, and also those formed due to the presence of an

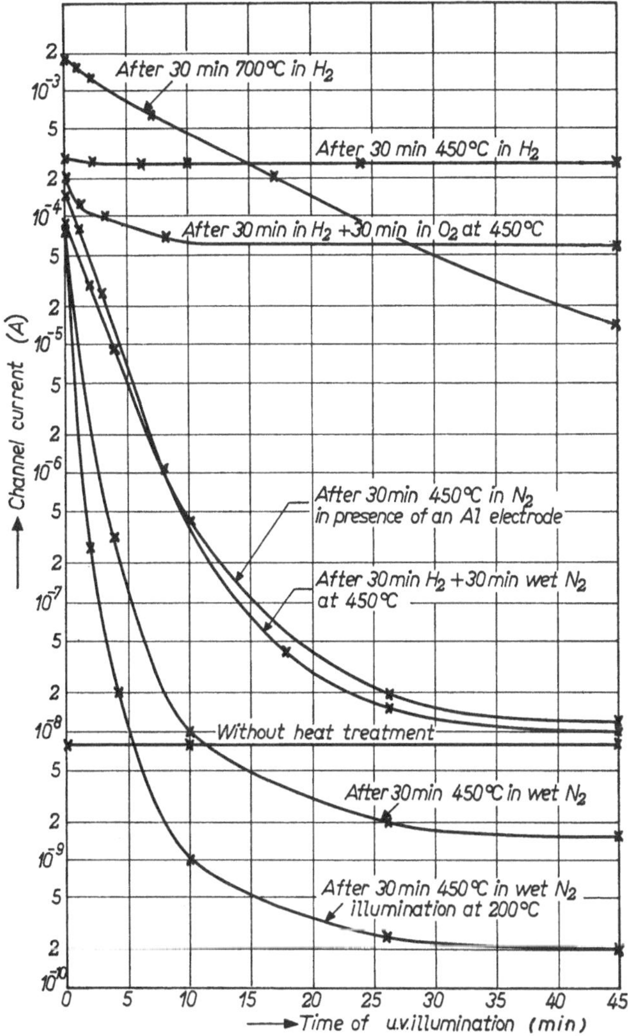

Fig. 6.10. Effect of irradiation by u.v. light on channel currents of samples which had been subjected to a phosphorus diffusion in N_2 and subsequently to various heat treatments. These heat treatments are indicated along the curves. In one case the sample was held at 200 °C instead of at room temperature while it was exposed to u.v. light.

aluminium electrode on the oxide during heating, although the effects were somewhat slower in these cases. However, the channels formed during heat treatment at 450 °C in hydrogen were hardly affected.

To find out whether the disappearance of the channels was due to a charge redistribution or (also) related to the formation of surface states in which the electrons of the inversion layer became trapped, we made MOS transistors in a few cases, both before and after irradiation, by deposition of aluminium electrodes. Both I_D-V_G and C-V_G measurements (to find the gate voltage needed to get "flat-band" conditions) were then carried out, results of which have been plotted in table 6-I. This table shows that the characteristics of the sample

TABLE 6-I

Effect of u.v. illumination on the surface properties of oxidized silicon, in relation to low-temperature heat treatments after the phosphorus diffusion. Measurements were carried out on MOS-transistor structures.
Sample A: heat treated in wet N_2 during 30 minutes at 450 °C.
Sample B: heat treated in dry H_2 during 30 minutes at 450 °C

		gate voltage at which:					
		$y_s = 0$ (V_f)	$I_D =$ 1 μA	$I_D =$ 10 μA	$I_D =$ 100 μA	$I_D =$ 1 mA	$I_D =$ 25 mA
A	before irradiation	−12	−6	−4	2	20	34
A	after 10 min exposure to u.v. light	−18	4	7	14	35	52
B	before irradiation	−16	−10	−8	−2	15	30
B	after 10 min exposure to u.v. light	−16·5	−10	−8	−2	15	30

which had been heat treated in hydrogen were scarcely affected. The C-V_G curve of the sample that had been heated in wet N_2 showed a shift to more negative voltages, whereas more positive gate voltages were necessary to induce the given values of the drain currents. It is thus concluded that u.v. light causes formation of electron-trapping centres. The presence of these centres causes possibly also a lower electron mobility in the inversion layer, so that the transconductance of the MOS transistor may have become lower for this reason too.

The electron-trapping centres which form during irradiation by u.v. light are not necessarily the same as those which disappeared during the preceding heat treatment. The radiation-induced defects are presumably related to the structure of the oxide film. We did a few other experiments to find out something more about that relation. A sample was first heated in H_2 at 450 °C for 30 minutes, so that the electron-trapping centres disappeared. After this heat treatment a heat treatment at the same temperature was carried out in wet N_2. It appeared that the properties of the surfaces in relation to the u.v. illumination were about the same as in the case where no preceding heat treatment had been given (fig. 6.10). In another case we heated first in wet N_2, afterwards in dry H_2. Ultra-violet light caused again a disappearance of the channels, as if the treatment in H_2 had not been given. It may therefore be concluded that in the type of oxidized surfaces considered, water vapour as well as hydrogen causes the disappearance of electron-trapping centres, but that water vapour at 450 °C acts also in such a way on the oxide film that electron-trapping surface states can be formed easily during illumination with u.v. light. A heat treatment in O_2 after a treatment in H_2 also had some effect in this respect (fig. 6.10), while heat treatment in N_2 instead of O_2 had not. Heat treatment in H_2 at higher temperatures resulted also in channel formation, but the effect of u.v. light increased. In fig. 6.10 an example of illumination has been given for a sample which had been heated in H_2 at 700 °C during 30 minutes. As the vertical scale is logarithmic in this figure, the effect is in fact very large in this case.

In contradistinction to the effects of X-ray irradiations, we have not found any influence of the removal of the phosphate glass from the top of the oxide on the results of illumination by u.v. light.

6.3.2.4. Alternating γ-ray or X-ray and u.v. irradiations

In table 6-II a few results of alternating irradiations have been given. The general effect was again that the channel currents increased during X-ray irradiation, but decreased during u.v. illumination. The u.v. illumination had a larger effect when first an X-ray or γ-ray irradiation had been given. The effect occurred very quickly, especially after the γ-ray irradiation and after a long period of X-ray irradiation.

The marked sensitivity for u.v. light in the latter cases allowed us to observe the changes in the drain current as a function of wavelength, using monochromatic light (which was of much lower intensity than in the other u.v.-illumination experiments). We could not observe any effect for wavelengths higher than 2950 Å. This corresponds to a threshold energy of 4·2 eV for the transport of electrons from silicon to oxide. With light of the same intensity the effect became larger when the energy of the quanta increased (measurements were done up to 6 eV).

6.3.2.5. Thermal annealing of irradiation effects

Not much attention was paid to this aspect, although it may well be that a profound study of these effects may give information about the structure of SiO_2 and its relation to the surface properties of the underlying silicon. Results of thermal annealing on five samples have been plotted in fig. 6.11. These samples had been made by the same oxidation and phosphorus diffusion as described before. Three of them were reheated at 450 °C in wet N_2 so that channels formed. In one of these samples the channel current was further increased by X-ray irradiation for 1 hour (150 kV, exposure rate 10^4 R/min), while another sample was illuminated by a low-pressure-mercury lamp for 30 minutes, so that the *n*-type channel almost disappeared. The results of heating in the temperature range 200-500 °C in dry N_2 (fig. 6.11) show that the irradiation

TABLE 6-II

Channel currents in planar *npn* structures after successive treatments. The oxide was covered by a phosphate glass. Irradiations were carried out with X-rays, γ-rays and u.v. light

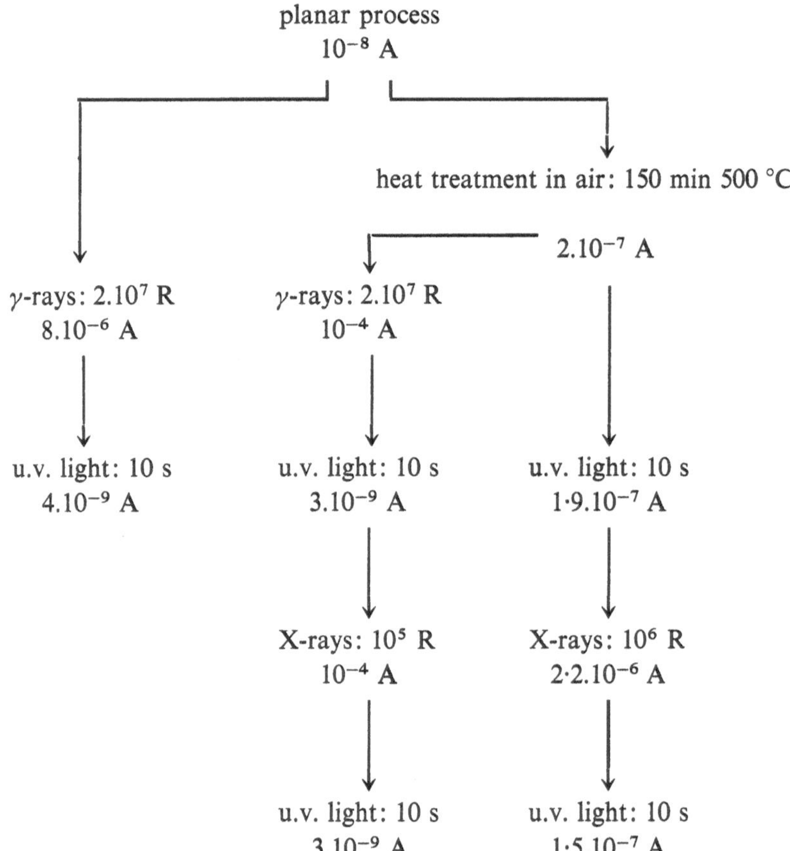

effects were fairly stable during heating at 200 °C for 1 hour, but considerable changes were observed at 300 °C, and at 400-500 °C the irradiation effects were soon annealed. It is interesting to note that the low channel current which was obtained during u.v. illumination became even smaller during heating at 200 °C. In fig. 6.10 it has been shown that the u.v. light could become more effective when the sample was held at 200 °C during illumination.

In fig. 6.11 the effect of thermal annealing has been shown too for samples that had not been subjected to a heat treatment in a wet ambient after the phosphorus diffusion. As was discussed before, the main effect of an X-ray irradiation on such a sample can be considered to be a charge redistribution so that a more effective positive charge is incorporated in the oxide. Again, heat treat-

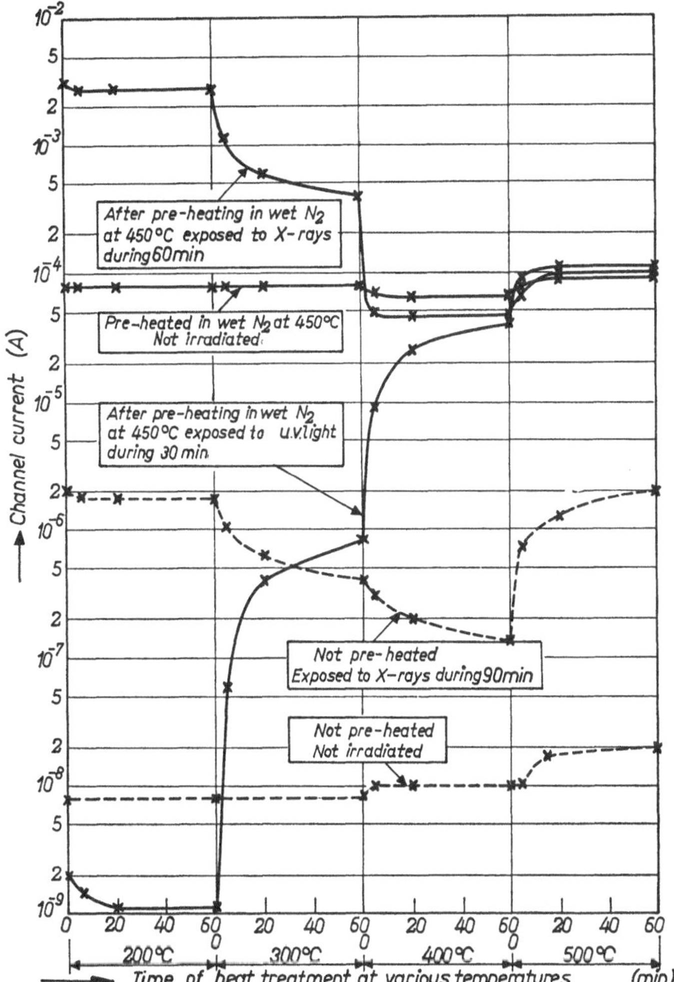

Fig. 6.11. Thermal annealing in dry N_2 of irradiation effects, induced in planar *npn* structures. The method of irradiation has been indicated at the various curves. For comparison the influence of the annealing treatment on non-irradiated samples is shown.

ment at 200 °C was not sufficient to anneal this effect, but this tendency became stronger at 300 and 400 °C. The increase of the channel current during heat treatment at 500 °C may be due to the effect of some moisture adsorbed at the surface before heating. As was discussed in chapter 4, such a presence of water may cause irreproducible results of heat treatments.

6.4. Discussion

6.4.1. *General effects of ionizing irradiation on solids*

Ionizing radiations are used frequently to observe changes in the light-absorption or paramagnetic-resonance spectra of various materials. In many cases the aim of such experiments is to correlate the observed phenomena with the structure of the material. The primary effect of an ionizing irradiation is a removal of electrons from ions in the network. The free electrons (or holes) may then become trapped somewhere else. In the new environments they may be able to interact with photons of certain energies and thus give rise to formation of new optical-absorption bands. The electrons may be removed again from the trapping centres by addition of suitable photon or thermal energy. These processes are often indicated as optical and thermal bleaching, respectively.

In addition to a redistribution of electrons over the various centres, irradiation may result in structural changes. In the cases of irradiations with particles of high energy, such as neutrons, this may be caused by direct displacement of atoms or ions. In the case of γ-rays, X-rays or u.v. light, new centres may form because chemical bonds may become weaker as a consequence of the removal of an electron and certain ions or atoms may then be able to take a new position.

6.4.2. *The structure of* SiO_2 *and its relation to effects of ionizing irradiations*

A number of attempts have been made to learn more of the structure of fused silica and quartz from irradiation experiments [6-8,9]). Although it seems worth trying to correlate these experiments with the results of irradiation on the Si-SiO_2 system, one should realize that there are a few circumstances which make the relationship difficult. In the case of ionizing irradiation on oxidized silicon electrons may be excited in the silicon as well as in the oxide and both have to be considered in the charge redistribution. Electrons which are excited in the oxide may be captured in trapping centres in the oxide but may also find their way to the silicon. Various experiments indicate that a considerable energy is needed to transfer these electrons again into the oxide. Another complication is that during irradiation of oxidized silicon Compton electrons, formed in the silicon, may induce an increased irradiation effect in the oxide. As on the average these Compton electrons have the same direction as the X- or γ-rays, a difference may be found depending on the direction of the radiation. We have observed that in many cases the effects of X-ray irradiation were somewhat larger when the radiation had to pass the silicon before entering the oxide.

Although the presence of the silicon near the oxide influences the effect of irradiation on SiO_2, on the other hand the semiconductor properties of the silicon may allow more information to be obtained about irradiation of SiO_2 in relation to its structure. The surface conductance of the silicon is related to this structure, but also to the structure of the silicon near the interface. During irradiation new centres may form both in the silicon and in the oxide, which make an explanation of the experimental results even more difficult.

6.4.2.1. Imperfections in SiO_2

Systematics of imperfections in silicon-oxygen networks have been given by Stevels and Kats [6–6]). They based their considerations on a simple model in which the Si^{4+} ions are bound together via "bridging" O^{2-} ions. Oxygen vacancies as well as interstitial oxygen ions were considered as most probable imperfections. In many cases the interstitial oxygen ions may be considered as "non-bridging" oxygen ions bound to only one silicon ion. In glasses another cation is often present near such a non-bridging oxygen ion, so that electro-neutrality is maintained. Hydroxyl groups may also be considered as such combinations. They may form in high concentrations in SiO_2 during heating in an ambient containing water or hydrogen [6–7]). In both processes it may be supposed that Si-O-Si bonds are broken up. However, in the case of heating in hydrogen the SiO_2 is also chemically reduced. In the following hypothetica-reaction equations this is indicated by converting an Si^{4+} ion to an Si^{3+} ion:

$$Si^{4+}\text{-O-}Si^{4+} + H_2O \rightarrow Si^{4+}\text{-OH} + \text{HO-}Si^{4+}, \tag{6.1}$$

$$Si^{4+}\text{-O-}Si^{4+} + \tfrac{1}{2} H_2 \rightarrow Si^{4+}\text{-OH} + \quad Si^{3+}. \tag{6.2}$$

At relatively low temperatures (below about 500 °C) a large amount of hydrogen may be dissolved as hydrogen molecules instead of as hydroxyl groups.

Except on interstitial places foreign cations may also take the place of silicon in the lattice. Elements like phosphorus and boron (which are frequently used in silicon-transistor technology) may be built in substitutionally. When high concentrations of foreign elements are present, the glassy structure may become completely different.

6.4.2.2. Effects of ionizing irradiations on fused silica

At this moment no clear information is available on the effects of irradiation on SiO_2 in relation to its structure, deviations from stoichiometry or impurity content. Many of the publications referred to [6–8-15]) indicate that the effect of irradiation on the optical-absorption and paramagnetic-resonance spectra of fused silica depends very much on the way in which the samples have been prepared. Before irradiation fused silica does not show much optical absorption down to wavelengths of about 1800 Å *) (corresponding to a photon energy of about 7 eV). Certain correlations have been found between the internal structure or impurity content and the radiation effects. In particular the presence of hydroxyl groups appeared to have a marked influence. X-ray irradiation on water-containing fused silica at a low temperature (77 °K) has resulted in paramagnetic-resonance spectra which could be explained as being due to the for

*) Fused silica with a low oxygen content may show optical absorption near 2400 Å, but there are indications that this occurs only if the silica is contaminated by germanium or aluminium [6–9]).

mation of hydrogen atoms [6-12]. It has been suggested that during irradiations at higher temperatures these atoms cannot exist for a long time. They may move so easily that they are able to combine with other centres.

In a study on the relation between certain colour centres and hydroxyl groups, Weeks and Lell [6-15] came to the conclusion that the SiO_2 network contains often a number of defects which are of a more complex nature than is accounted for by a simple model with single oxygen vacancies or interstitials.

Nelson and Crawford [6-13], supported by Arnold and Compton [6-11], have suggested that in fused silica certain new centres may form during ionizing irradiations as a result of the rupture of strained covalent bonds. The absence of strained bonds in crystalline SiO_2 could possibly explain that these centres do not form during irradiation of quartz. This possibility for formation of new centres may be of importance in irradiation of oxidized silicon. Except the strain due to the fused-silica-like structure of the SiO_2 in these cases, other reasons for strain are present because of the difference of expansion coefficients and the mismatch between the structures of SiO_2 and Si.

Because X-ray as well as u.v. irradiation was employed in our investigations, it is interesting to note that as regards the formation of colour centres in silicate glasses both kinds of irradiation have to a certain extent the same influence [6-10]. On the other hand, it has been shown that absorption bands induced by γ-rays can be bleached by u.v. light [6-14]. Thermal bleaching occurs fairly quickly at temperatures above 300 °C.

6.4.2.3. The structure of oxide films grown thermally on silicon

The oxide layers which form on silicon during thermal oxidation have dielectric properties, an index of refraction and a specific density which are comparable to those of fused silica. It seems reasonable therefore to assume that their structure consists also of a network of SiO_4 tetrahedra. Imperfections may be present which are comparable to those which can occur in fused silica. However, the number and the nature of the imperfections will vary across the film. Near the silicon surface an increased amount of defects has to be present to account for the mismatch between the silicon lattice and the Si-O network. Moreover, deviations from stoichiometry can vary across the film. An oxygen-concentration gradient will be frozen in during cooling after heat treatment in an oxidizing atmosphere. During etching, preoxidation cleaning treatments and oxidation a number of impurities may be adsorbed at the surface and become incorporated in the oxide film during heat treatment. Donor and acceptor elements present in the silicon can be taken up in the oxide layers under certain circumstances. Otherwise they may be introduced during experiments in which oxide films are used as a masking agent against diffusion of donors and acceptors into silicon. When hydrogen or water has been present during heat treatments, a number of hydroxyl groups may have been built in.

In chapter 4 it was demonstrated that heat treatment in the presence of water or hydrogen had considerable influence on the surface properties of the oxidized silicon. The now presented experiments indicate that the presence of hydrogen might also have considerable influence on the irradiation effects. In the discussion of the results we will mainly concentrate on this aspect.

6.4.3. *Discussion of the experimental results*

It has been shown in chapter 5 that certain effects of ionizing irradiations in MOS transistors can be related to charge redistributions. In those experiments the results were found to depend very much on the voltage which was applied across the system during irradiation, and the charge-transfer processes at the M-O and the O-S boundary had a marked influence on the results. The experimental results given in this chapter are mainly concerned with effects at the O-S boundary, as in most cases no metal electrode was present and no external bias was applied during the irradiations.

The results will be discussed further in terms of charge redistributions and the formation of new centres in the oxide-silicon system.

6.4.3.1. Charge redistributions in the oxide-silicon system

The results indicate that ionizing irradiations, whose photon energy exceeds 4·2 eV tend to cause a redistribution of charge between the oxide and the silicon substrate. Both the measurement of channel currents and the capacitance measurements show that the primary effect of X-rays was in general such that electrons were transported from the oxide to the silicon, leaving positive charge in the oxide. Ultra-violet light tended to cause the opposite effect, thus electron transport from silicon to oxide. Figure 6.6 shows one example (a sample oxidized in wet O_2 at 1100 °C for 1 hour) in which the primary effect of u.v. light was of the same sign as that of X-rays, that is to say an increase in channel current. Note that a comparable sample showed a very large increase in channel current during X-ray irradiation (fig. 6.2).

Obviously the irradiation effects depend strongly on the method of oxide preparation. The cooling procedure too may have a considerable influence on the defect and charge distribution which has been frozen in. We point to the fact that even temperature gradients during heat treatment may cause certain charge redistributions, which can also affect the behaviour of a sample during irradiation [6-16]). Therefore it is not readily possible to relate the irradiation effects quantitatively to the oxidation process. Nevertheless a general distinction can be made between the behaviour of "wet" and "dry" oxides during irradiation. The small effects of irradiation on the channel currents of samples made in a dry ambient are certainly related to the relatively large amount of electron-trapping centres at the surface in these cases. The presence of such centres causes the channel current to be rather insensitive to charge redistribution. A comparison of figs 6.3 and 6,5 shows that the *C-V* curve shifted over 40 volts during X-ray irradiation, while the channel current increased only from 0·05 to 1·1 μA in this sample, which had been reoxidized for 16 hours at 1200 °C (oxide thickness 0·8 μ). We noticed that X-ray irradiation tended to have an increasing influence on samples with "dry" oxide layers, when these films were increased in thickness, thus made at higher temperature or (and) during a longer oxidation period. This effect is probably related to changes in the structure of the oxide during growth. It seems possible that in these cases more experiments may disclose certain relations between oxide structure, irradiation effects and surface properties of the silicon.

We also point to the remarkable influence of a phosphate glass on top of the SiO_2 film during X-ray irradiation. The presence of this glassy layer influenced the charge distribution, but not in the way that a part of the positive surface charge was present in this layer; the channel current remained almost equal after removal of the glass (fig. 6.9)! However, it decreased during further irradiation, until the sample showed about the same properties as the sample from which the phosphate glass had been removed at the early beginning of irradiation. An explanation may be found if one assumes that there is still some external electric connection present during irradiation, caused by leakage across the surface of the oxide film, or, if present, through the phosphate glass. A difference in the photo-electromotive force in the glass-oxide-silicon structure compared to the oxide-silicon structure (glass removed) would then cause a different charge distribution at the O-S boundary.

6.4.3.2. Radiation-induced defects

A number of experiments point to the formation of new centres during irradiations. Especially in the cases of X-ray and γ-ray irradiations their formation leads to a decrease in the transconductance of MOS transistors or a decrease in the slope of the high-frequency *C-V* curve. Hysteresis effects in these curves as well as their dependence on frequency point to formation of new surface states. Their formation is obviously much more pronounced in "wet" than in "dry" oxides, because the slope of the *C-V* curve of a sample with a "wet" oxide decreased considerably during irradiation (compare fig. 6.4 with fig. 6.5). In fig. 6.2 it was shown that the channel currents of samples covered by a "wet" oxide primarily showed an increase followed again by a decrease. In sec. 6.4.3.1 the increase was related to an electron flow from oxide to the silicon, which corresponds also with the shift of the *C-V* curve to more negative voltages. The decrease in channel current, which occurred during continued irradiation has to be related to the formation of new centres, in which the electrons become trapped instead of the silicon (the inversion layer) itself.

Obviously the formation of these new centres was not very pronounced in dry oxides. A decrease in channel current was still sometimes found after the first increase (fig. 6.3), but in many other cases continued irradiation gave a slight increase in channel current, which may be due to formation of new donor states. One should realize, however, that the samples covered by a "dry" oxide exhibited already a large number of electron-trapping centres at the surface, which makes it more difficult to observe the formation of new centres. The same can be said about the samples with an oxide covered by a phosphate glass. Here both the C-V curve and the I_D-V_G curve shifted over about the same voltage difference during X-ray irradiation (fig. 6.8), although this was not always true for samples in which the electron-trapping centres had been decreased by low-temperature heat treatments. As water or hydrogen often played an important role in these heat treatments too, it is reasonable to suggest that the formation of new electron-trapping centres at oxidized surfaces during irradiation is related to the presence of defects such as hydroxyl groups in the oxide.

As discussed before, u.v. light with a photon energy of at least 4·2 eV can be considered to be able to excite electrons in the silicon to a sufficiently high energy level that they may enter the oxide film. When positive charge is present in the oxide, this transport will be aided by the electric field at the oxide-silicon interface. However, this transport can only be effective in the disappearance of inversion layers at the surface of p-type silicon, when centres are present in the oxide film in which the electrons may become trapped. Obviously a sufficient number of such trapping centres was not always present, so that the effect of u.v. light was often very small.

The samples covered by a "dry" oxide showed rather small inversion layers. However, also in these samples there have to be positively charged centres either in the oxide or in the surface region of the silicon to account for the negative surface charge further present. As u.v. light had only very small effects on these samples, one may conclude that these positive centres do not easily capture an electron. They behave like donors, whose activation energy to donate an electron is rather small. It cannot be concluded whether these donor centres are present in the oxide or in the silicon. It has been shown (figs 6.3 and 6.5) that X-ray irradiation on samples with a "dry" oxide can cause a transport of electrons from the oxide to the silicon. The centres from which these electrons were removed cannot be considered as such donors, but rather as trapping centres, with a high activation energy to remove electrons. During illumination with u.v. light the X-ray effects were soon offset again, which may be interpreted as a return of the electrons to the centres from which they were excited during X-ray irradiations.

The samples with "wet" oxides sometimes showed an almost immediate decrease in channel current during illumination by u.v. light, as well as a shift of the C-V curve to a less negative voltage. It may therefore be concluded that in such oxides, besides the presence of positively charged donor centres such as occur in the samples with a dry oxide, also other centres are present, in which electrons may become trapped. There should be some relation between the oxidation method (vapour pressure of water in the gas and temperature and duration of oxidation) and the number of such defects, but due to somewhat irreproducible results of the oxidation experiments, we were not able to find this relation. It may be interesting to note that we have observed that n-type channels present at the surface of p-type samples covered by an oxide made by burning tetraethyloxy-silane in oxygen at 350-400 °C disappeared very quickly during irradiation by u.v. light. These types of oxide probably contain many more defects than those grown by thermal oxidation of silicon.

The effect of u.v. light on the transport of electrons from silicon to the oxide is thus dependent on the number of trapping centres in the oxide. As discussed before, the number of such centres may be increased during X-ray irradiation. This explains why the effect of u.v. light could be increased by irradiating the samples first with X-rays or γ-rays (fig. 6.7, table 6-II). However, u.v. light itself must also be considered to be able to induce new trapping centres in the oxide in some cases. This effect can again be neglected in "dry" oxides, but has a greater influence in "wet" oxides (fig. 6.6). Formation of new surface states during u.v. illumination is also clearly demonstrated in table 6-I (sample A).

There seem to be certain relations between the irradiation effects and properties of the oxide films, such as density, refractive index, and infrared-absorption spectra. Such measurements were carried out by Pliskin and Lehman [6-17]) and led to the conclusion that "dry" oxide films have a more dense structure than "wet" oxides, while pyrolytically grown films contain many more defects. Relating this knowledge to the effects of u.v. illumination one may conclude that channels disappear more rapidly during u.v. illuminations when the oxide contains more defects. Hydroxyl groups can be considered as imperfections which may be attacked by u.v. light. As they are probably present in most of the "less dense" oxides, they may play an important role in the formation of new centres during irradiation. In sec. 6.4.2 the possibility has been mentioned that hydrogen is removed from a hydroxyl group during irradiation. The remaining oxygen ion or the centre at which the hydrogen atom reacts may then be able to capture an electron*).

In sec. 6.4.2 it was stated that hydroxyl groups can form during reaction of water or hydrogen with SiO_2, although the reaction of hydrogen with SiO_2 is not very pronounced below 500 °C. In this relation the effect of u.v. light on samples which had been subjected to a phosphorus diffusion and subsequently to various treatments at low temperatures is interesting (fig. 6.10). The samples which had not been subjected to further heat treatment are not affected by u.v. illumina-

*) The oxides considered in this chapter probably contained traces of sodium. As the presence of this element may cause disorder in the structure of the oxide and its boundary to the silicon (see chapter 7), its presence may also be assumed to cause some excess irradiation effects.

tion. Also the effect of X-rays on such a sample (fig. 6.8) points to the fact that new centres are not easily formed in this oxide. In chapter 4 it has been suggested that the good quality of this oxide as well as the large amount of trapping centres at the surface are due to the gettering action of the phosphate glass on top of this oxide for water and possibly also for other impurities. Moreover, the oxide is presumably in a somewhat reduced state (the phosphorus diffusion was carried out in nitrogen) and may have a low content of hydroxyl for that reason too.

As was discussed in chapter 4, the formation of channels at the surface of these samples during heating in water vapour or hydrogen is to be explained as being mainly due to the disappearance of surface states, which can mask the influence of fixed charge in the oxide and also of the voltage across MOS structures made on such samples. Because of the fact that water vapour and hydrogen gave about the same result, it was thought that in both cases a reaction of unsaturated bonds wit hydrogen caused the disappearance of the surface states. The results given in sec. 6.3.2.3 (fig. 6.10, table 6-I) seem to indicate that the heat treatment in water vapour at 450 °C had moreover some destructive action (formation of hydroxyl groups?) on the oxide, while H_2 did not, because there was no effect of u.v. light in the latter case. Presumably H_2 could not attack the SiO_2 network at this temperature, but was able to react with dangling bonds near the oxide-silicon interface, thus forming covalent Si-H bonds, which are not so readily attacked by radiation as O-H bonds. Otherwise it might be possible that centres formed during reaction of H_2 and the SiO_2 structure are of a completely different nature than those formed during reaction with H_2O (sec. 6.4.2.1). However, the supposition that this is the reason of the observed differences seems to be ruled out by the fact that the sensitivity to u.v. light was also present when a treatment in water vapour was followed by a heat treatment in hydrogen or vice versa. Heating at 700 °C in H_2 resulted in strong inversion at the silicon surface, but the u.v. effect was very pronounced after this heat treatment, indicating that at this temperature H_2 was able to attack the SiO_2 network. Whether this does occur or not, probably depends also on the oxygen content of the oxide. Some H_2 may dissolve in molecular form and when a subsequent heat treatment in O_2 is given, the chance of formation of OH groups may be greater. This might explain the increased effect of u.v. light in such a case (fig. 6.10).

An aluminium electrode present during heat treatment at 450 °C in dry N_2 had about the same effect on the surface properties of oxidized silicon as heat treatment of an oxide-covered slice in wet N_2 and the u.v. effects are also comparable (fig. 6.10). This may be considered as a support for the supposition given in chapter 4, that the effect of the aluminium has to be considered as an increased effect of the presence of traces of water. Reduction of the water species (or OH groups in the oxide) by the aluminium may give the hydrogen needed for the annihilation of the surface states. However, the effect of the aluminium cannot be compared simply with the effect of heating in hydrogen. Although in both cases the surface states disappear, the effect of u.v. light is very different (fig. 6.10), indicating that probably the diffusing species, in the case of the aluminium being present, are hydrogen atoms, which can cause a saturation of the unsaturated bonds, but are also able to attack Si-O bonds.

6.5. Conclusions

Ionizing irradiations can cause charge redistributions in silicon–silicon-dioxide systems. During X-ray or γ-ray irradiation a number of free electrons may form in the oxide, which tend to flow to the silicon, thus leaving a positive charge in the oxide. A counteracting field is thus formed and the equilibrium situation is such that the silicon surface has become more n-type. However, the equilibrium may be affected during continued irradiation because new centres form either in the oxide or in the surface region of the silicon. In many cases the presence of hydroxyl in the oxide appears to promote the formation of new defects during irradiation. Consequently n-type channels present on p-type silicon oxidized in a wet ambient tend to disappear during continued X-ray or γ-ray irradiations.

Ultra-violet light with a photon energy of at least 4·2 eV can give a sufficient amount of energy to electrons in the silicon to make transfer to the oxide possible. Such a transport will only occur and result in a change in the surface potential of the silicon when the electrons excited from the silicon can be captured in the oxide. In oxides with a "dense" structure, such as those made in dry oxygen, this effect is almost absent, which leads to the conclusion that those positive ions which are responsible for the n-type surface in these cases, do not easily capture an electron. In some other types of samples, such as those made in a wet ambient or by pyrolysis of tetraethyl-oxy-silane, a sufficient amount of trapping centres is probably present to guarantee capture of electrons coming from the silicon during illumination with u.v. light. It is difficult to make out, however, whether these centres are already present or induced during irradiation.

As the effect of u.v. light on the disappearance of n-type channels depends very much on the presence of centres in the oxide in which the electrons can become trapped, its effect can be increased by giving first an X-ray or γ-ray irradiation. Although thus means are present to cause the disappearance of the channels commonly found after oxidation of p-type silicon, the practical use is limited because of the thermal instability of the surface properties after irradiation. At 300-500 °C thermal annealing is very effective.

Ionizing irradiation may be used for detecting differences in properties of oxide-covered silicon surfaces. In chapter 4 it has been shown that electron-trapping centres, present in large numbers (10^{12}-10^{13} per cm^2) on the surface of oxidized p-type silicon after this was subjected to a phosphorus diffusion in nitrogen, can disappear when they can react with hydrogen during heat treatment. The hydrogen may be present in molecular form in the ambient or result from a reaction of water with the oxide film. A similar effect is induced when during heating in a dry ambient an aluminium electrode is present on top of the oxide. The three methods yield about the same electrical properties as far as measurements on MOS structures are concerned. However, illumination with u.v. light after these treatments has almost no effect for the case of heat treatment in hydrogen gas (at 450 °C), but very large effects are found in the other two cases. In the latter two cases the results may be related to the presence of hydroxyl groups, which probably do not form in large amounts when molecular hydrogen diffuses through the oxide at low temperature.

REFERENCES

6-1) E. Kooi, Philips Res. Repts **20**, 306-314, 1965 (chapter 5 of this book).

6-2) H. L. Hughes and R. G. Giroux, Electronics **37**, 58-60, 1964.

6-3) H. Edagawa, Y. Morita, H. Ishikawa, S. Maekawa and Y. Inuishi, Jap. J. appl. Phys. **3**, 644-660, 1964.

6-4) H. Edagawa, Y. Morita and Y. Inuishi, J. phys. Soc. Japan **18**, 314-315, 1963.

6-5) M. V. Whelan, Philips Res. Repts **20**, 562-577, 1965.

6-6) J. M. Stevels and A. Kats, Philips Res. Repts **11**, 103-114, 1956.

6-7) A literature review on the diffusion of H_2O and H_2 in fused silica has been given by R. W. Lee, Phys. Chem. Glasses **5**, 35-43, 1964.

6-8) A number of irradiation effects in SiO_2 has been reviewed in J. H. Schulman and W. D. Compton, Color centers in solids, Pergamon Press, 1963, pp. 291-306.

6-9) A brief review on the colouring of fused silica in relation to its impurity content or method of preparation has been given by G. Hetherington, K. H. Jack and M. W. Ramsay, Phys. Chem. Glasses **6**, 6-15, 1965.

6-10) A. Kats and J. M. Stevels, Philips Res. Repts **11**, 115-156, 1956.

6-11) G. W. Arnold and W. D. Compton, Phys. Rev. **116**, 802-811, 1959.

6-12) J. S. van Wieringen and A. Kats, Archive des Sciences **12**, 203-204, 1959.

6-13) C. H. Nelson and J. H. Crawford Jr, J. Phys. Chem. Solids **13**, 296-305, 1960.

6-14) C. H. Nelson and R. A. Weeks, J. Am. ceram. Soc. **43**, 396-404, 1960.

6-15) R. A. Weeks and E. Lell, J. appl. Phys. **35**, 1932-1938, 1964.

6-16) E. Kooi and M. M. J. Schuurmans, Philips Res. Repts **20**, 315-319, 1965.

6-17) W. A. Pliskin and H. S. Lehman, Paper given at the Meeting of The Electrochemical Society in Washington, October 1964, Electronics Division, abstract 128.

7. THE SURFACE CHARGE IN OXIDIZED SILICON*)

Abstract

Investigations carried out to study the influence of oxidation and further heat treatments on the surface properties of silicon suggest that hydrogen and sodium impurities have a considerable influence. The presence of sodium was checked by neutron-activation analysis, the influence of hydrogen was investigated by heat treatments in a wet ambient. A model is proposed which explains qualitatively why the presence of sodium causes the formation of positive surface charge during heat treatment and why at low temperatures hydrogen (or water) can cause a decrease of this charge as well as a decrease in the number of surface states. Further experiments indicate that the sodium content of SiO_2 films can be decreased considerably by heat treatments under conditions of low oxygen pressure.

7.1. Introduction

The coating of silicon surfaces by thermally grown silicon-dioxide films has already been the subject of a large number of investigations, many of which indicate that the resulting surface properties depend on various processing factors. For example, a distribution of donors and/or acceptor impurities between the silicon and the oxide may occur during oxidation. These effects tend to make the surface more *n*-type or less *p*-type than the bulk of the crystal. In practice, however, the surfaces are in many instances much more *n*-type than may be expected from these effects. This can be considered as being due to the presence of positive charge in the oxide film and/or a predominance of donor-type surface states to be related to centres in an interface region at the Si-SiO_2 boundary.

The presence of surface states can be indicated by field-effect measurements, in which one may make use of MOS (metal-oxide-semiconductor) structures. Both differential MOS capacitance (C) versus d.c. voltage (V) and, in MOS transistors, drain-current (I_D) versus voltage measurements are suitable methods. However, it is difficult to distinguish charge in interface states near the valence and conduction bands from (fixed) charge in the oxide film, as they both cause a displacement of the I_D-V and C-V curves along the voltage axis. The slope of the C-V curves may be affected by surface states distributed through the energy gap. Whether such states are present or not, the voltage V_f at which the MOS capacitance corresponds to flat-band conditions may be used as a reference point. From V_f a number (N_f) of surface charges per cm^2 at the Si-SiO_2 interface may be calculated (approximately: $eN_f = C_{ox} V_f$, $e =$ unit charge, $C_{ox} =$ oxide capacitance per cm^2). Values of V_f of the order of 1 V or less may be explained by a work-function difference between the metal electrode

*) Published: Philips Res. Re;t; **21**, 477-495, 1966.

and the silicon. When charge is distributed through the oxide only an effective part of it is reflected in the value of N_f. In general, however, the (positive) surface charge in oxidized silicon is located closely to the Si-SiO_2 interface as a large part of the oxide film may be etched away before N_f is affected.

Changes in N_f may occur when ions are able to drift in the oxide film, for example when a bias is applied across the MOS structure. Considerable instability effects can be observed when the drift experiments are carried out at elevated temperatures (above about 100 °C) and the metal of the MOS structure is made the positive electrode. Snow et al. [7-1]) have suggested that these drift effects are due to the presence of alkali-metal impurities, especially sodium. Indeed it has appeared [7-2,3]), and will be confirmed in this chapter, that after a usual oxide preparation the top layer of an oxide film may contain a fairly large concentration of sodium ions. Snow et al. [7-4]) report also that carefully prepared oxide films do not show such instability effects and that the N_f values in MOS structures with oxide films of this type are low (in the range $1\text{-}3.10^{11}$ cm^{-2}, for oxidation at 1200 °C in oxygen). This suggests that excess surface charge is due to (sodium) contamination. Indeed, it has been found that the *n*-type character of oxide-covered silicon surfaces can be increased by heat treatment in a sodium-containing environment [7-2,5]).

In a publication of Revesz and Zaininger [7-6]) a relationship was shown to be present between the oxidation rate at the end of oxidation processes in oxygen and the resulting N_f values. The thicker the oxide film, the lower the oxidation rate and the lower the resulting N_f, with a minimum of about 10^{12} cm^{-2}. These authors relate these effects to disorder in the structure of the Si-SiO_2 boundary, which will have more time to become ordered when the oxidation rate decreases. In accordance with this model the surface charge could be decreased by heat treatment of oxidized samples in an inert gas. Further evidence for the influence of the interface structure is given by the observations that the surface orientation of the oxidized crystal can have a marked influence on N_f[7-7-9]) and on the number of surface states [7-10,11]).

Effects of surface orientation and oxidation rate on the N_f values are confirmed in the work presented here, but it proves to be possible to relate these effects to the presence of sodium at the Si-SiO_2 interface. However, apart from alkali ions hydrogen is also known to be able to induce surface charge, although the role of this element is more intricate. In earlier work [7-12,13]) it was shown that the number of surface states can be decreased by low-temperature treatments (below about 600 °C) in the presence of hydrogen or water, whereas subsequently a slight decrease of N_f was often noted. At more elevated temperatures hydrogen treatments [7-13]) as well as treatment in water vapour [7-11]) cause less decrease or even an increase of N_f. In this chapter a model will be proposed which describes in a qualitative manner the relationship between N_f and various experimental conditions.

7.2. Sample preparation and measurements

Silicon slices of both ⟨111⟩ and ⟨100⟩ surface orientation were cut from 5-Ωcm *p*-type (boron-doped) and *n*-type (phosphorus-doped) crystals, made by the floating-zone method. After lapping with fine abrasive powder they were etched in an aqueous solution of HF (50%) and $KMnO_4$ (4%) so that at least a top layer of 50 microns was removed. The samples were then washed in distilled water and treated in hot nitric acid before oxidation.

Oxidation was carried out in a heated tube of fused silica, either in dry (boiling liquid) oxygen or in wet oxygen (oxygen bubbled through purified water). In some experiments the "dry" and "wet" oxygen were mixed to study the influence of the water content of the oxidizing gas, which was checked at the end of the furnace tube by dew-point measurements.

Oxides covered by a phosphate glass were prepared by subjecting oxidized samples to a P_2O_5 diffusion from a mixed SiO_2-P_2O_5 source for 30 minutes at 1050 °C in N_2. The composition of the phosphate glass formed in this way was about 7 SiO_2. P_2O_5, its thickness approximately 0·1 micron.

For MOS-capacitance measurements aluminium spots with a diameter of 0·5 mm were deposited on the oxide film. The silicon substrate was not heated during the deposition; *C-V* measurements were done with a measuring frequency of 500 kc/s. In the calculations to determine V_f, the work-function difference between aluminium and *n*-type silicon was assumed to be zero. The N_f values given in various tables are average values measured on one or more samples. The spread in results was always less than ± 10% of the indicated numbers.

The neutron-activation analysis of oxidized samples was carried out after irradiating them together with a sodium standard for 3 hours in a neutron flux of 10^{14} neutrons/cm^2/s *). The reaction Na^{23} (n, γ) Na^{24} results in activity of Na^{24} (half life 15·4 hours). Sections of the samples were dissolved in suitable buffered HF solutions, so that the sodium concentrations throughout the oxide films could be determined. For this purpose the solutions of the various sections were placed in a well-type scintillation detector coupled with a gamma spectrometer. Measurements were done on the 1·38-MeV peak of the gamma spectrum of Na^{24}. In order to increase the efficiency of removing radioactive sodium from the samples, all etch solutions used in the analysis were contaminated with non-irradiated sodium chloride.

7.3. Experimental results

7.3.1. *Surface charge in the presence of sodium contamination*

7.3.1.1. Influence of oxide thickness

In table 7-I a number of results have been summarized for *p*-type samples of

*) The irradiations were done in the reactor 2 at the Centre d'Etudes de l'Energie Nucléaire at Mol, Belgium.

TABLE 7-I

Influence of oxidation time and further treatments on N_f values (from the "flat-band" point in 500-kc/s MOS *C-V* curves) of *p*-type Si with ⟨111⟩ and ⟨100⟩ surface orientation. For comparison numbers have also been given for ⟨111⟩ *n*-type samples with an oxide thickness of 0·2 micron. Sodium concentrations are given for a top layer of 100 Å (total Na per cm^2) and in the remaining part of the films (average per cm^3)

oxidation period (h)	after oxidation (O_2 1200 C°)					N_f ($cm^{-2} \times 10^{11}$) after various treatments					
	oxide thickness (μm)	Na in SiO_2		N_f ($cm^{-2} \times 10^{11}$)		N_2-H_2O 30 m 450 °C		P_2O_5 30 m 1050 °C		P_2O_5 (1050 °C) + N_2-H_2O (450 °C)	
		top 100 Å (cm^{-2})	rest (cm^{-3})	⟨111⟩	⟨100⟩	⟨111⟩	⟨100⟩	⟨111⟩	⟨100⟩	⟨111⟩	⟨100⟩
$\frac{1}{4}$	0·1			28	11	10	7	15	3	10	3
1	0·2	5.10^{12}-10^{13}	2.10^{17}-5.10^{17}	15 (*n* : 11)	6	7 (*n* : 5)	3	7 (*n* : 2)	3	4·5 (*n* : 2)	3
4	0·4			11	5	3·5	1·5	2	4	1·5	2·5
16	0·8	5.10^{12}-10^{13}	5.10^{16}-10^{17}	9	2·5	2·5	0·5	6	2	2	2

both ⟨111⟩ and ⟨100⟩ surface orientation, oxidized for various periods at 1200 °C in oxygen (dew point —40 °C). The oxidation time is seen to have a considerable influence on N_f and the values for the ⟨111⟩ samples correspond closely to those found by Revesz and Zaininger [7-6]). The ⟨100⟩ surface-oriented samples, oxidized at the same time as the ⟨111⟩ samples show lower values of N_f, but in these samples, too, the lowest values are found for the thickest oxides. Comparison of *n*-type and *p*-type ⟨111⟩ samples shows lower N_f values for the *n*-type material.

In a number of samples the sodium contamination was measured. It appeared that a considerable amount of sodium was present on the top of the oxide (5.10^{13} to 10^{14} atoms per cm^2), but the major part of it could be washed off in water. It is difficult to make out which part of this sodium contamination is caused by the oxidation process. Samples from which the top layer (100 Å) had been removed before the irradiation showed afterwards the same sodium contamination on the outer oxide surface. In the results henceforth presented we will therefore pay no attention to the sodium that could be washed off in water. An example of the sodium distribution in an oxidized sample is shown in fig. 7.1 for

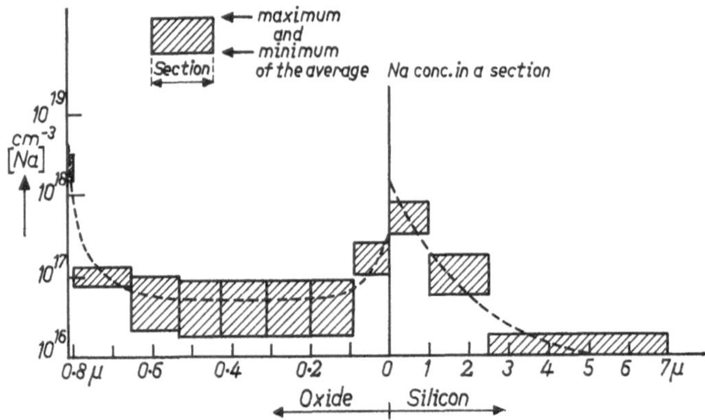

Fig. 7.1. Distribution of sodium in an oxidized ⟨111⟩ surface-oriented *p*-type silicon sample, found by neutron-activation analysis employing subsequent sectioning of the oxide film and the silicon substrate.

an oxide thickness of 0·8 micron. Although the profile could not be determined very accurately, there is a distinct accumulation of sodium near the top of the oxide films and at the Si-SiO_2 boundary, in the latter case in the oxide as well as in the silicon. In comparing various samples, we found the amount of sodium in the silicon to be much less reproducible than in the oxide film. Probably the sodium is not distributed homogeneously along the surface. It has been suggested [7-2]) that the pile up of sodium in the oxide near the interface is responsible for the surface charge. Qualitatively such a relationship is certainly present. In table 7-I a number of average sodium concentrations in the top layer (100 Å) and the remaining part of the oxide films have been given. The indicated numbers correspond to results found for two to four samples with oxide thicknesses of 0·2 and 0·8 micron. The sodium concentrations in the interior of the oxide film are lowest for the thickest films. The spread in the results made it impossible to detect whether there was any difference in sodium-concentration profile in the oxides made on ⟨111⟩ and ⟨100⟩ surface-oriented samples.

7.3.1.2. Influence of a P_2O_5 diffusion

A gettering effect of the phosphate glass for sodium, reported earlier [7-2,3]), was confirmed. However, the neutron-activation analysis showed that the gettering action did not remove all the sodium from the underlying SiO_2. Probably there was some sodium profile in the SiO_2 film in these cases too, but the sodium concentrations were too low to detect it accurately. Assuming that the amount of sodium at the Si-SiO_2 interface is indeed of large influence on the surface charge, the presence of such a profile would cause the surface charge to depend on the distance between the Si-SiO_2 and the phosphate glass-SiO_2 interface. This was in fact found, as can be seen in table 7-I: when samples with oxide films of varying thickness were subjected to the same P_2O_5 diffusion the largest amounts of charge were found for the thinnest films.

7.3.1.3. Influence of water vapour during heat treatment

It is apparent from table 7-I that the surface charge can also be lowered when the oxidized samples are treated in wet nitrogen at 450 °C. This effect is still present (although less) for samples covered with a phosphate glass. A possible explanation will be given later (sec. 7.4).

The gas used for preparing the samples mentioned in the previous sections was not extremely dry. To investigate the influence of the presence of water during oxidation we made oxide films of approximately 0·2 micron thickness by oxidation in oxygen with varying water content, using further the same oxidation system as used for the samples of table 7-I. Together with (in two cases) the average sodium concentration in the oxide film, the results are shown in table 7-II. The surface charge is largest for the cases that the gas was very dry, but for the higher water contents, too, an increase is observed. A tendency for the amount of sodium to decrease when the water content was increased can be noted.

7.3.2. *Oxidation under conditions of little sodium contamination*

The results of table 7-I show that the surface concentration of sodium in the oxide film is the same for oxidation during 1 hour and 16 hours. This indicates that in these cases the gaseous environment during oxidation contains a fairly constant amount of sodium impurities. Alkali ions which are adsorbed during etching or further "cleaning" treatments would probably tend to disappear during heating if the gaseous ambient were free of sodium. It is felt that the main source of sodium is due to contamination of the vitreous silica tube in which the oxidation was carried out. Particularly contamination of its outer surface (e.g. dust) may have been the predominant source because the alkali ions move easily through SiO_2 at elevated temperatures. However, the oxidizing gas (e.g. not sufficiently purified water) and drying agents for it can also be sources of contamination.

TABLE 7-II

N_f values, determined from *C-V* curves measured at 500 kc/s, before and after a treatment in wet N_2 at 450 °C (30 min), for ⟨111⟩ 5-Ωcm *p*-type silicon samples oxidized at 1200 °C in oxygen of different humidity. The oxide thickness is in each case approximately 0·2 micron. Sodium concentrations were measured on two series of samples made under different conditions

dew point of O_2 (°C)	oxida-tion time (min)	Na in SiO_2		N_f ($cm^{-2} \times 10^{11}$)	
		top 100 Å (cm^{-2})	rest (cm^{-3})	after oxidation	oxidation + wet N_2
−70	60			21	7
−40	60	5.10^{12}-10^{13}	2.10^{17}-5.10^{17}	15	7
−34	60			14	7
−20	60			11	5
− 6	60			10	6
25	30	2.10^{12}	4.10^{16}	7	6
65	12			9	6
90	7			15	8

During further experiments we succeeded in making oxide films with minor sodium contamination. This is indicated in table 7-III by some measurements of the sodium concentrations, which were at the limit of detectability. As the detected amounts were only of the order of 1% of the sodium contamination on the top of the oxide, which could be washed off in water, it is not even certain whether or not the detected concentrations are due to some residues of contamination induced during handling and irradiation of the oxidized slice. Table 7-III also presents N_f values of *p*- and *n*-type material of both ⟨111⟩ and ⟨100⟩ surface orientation. A comparison with table 7-I shows that the decrease of sodium contamination is accompanied by lower amounts of surface charge. The influences of oxide thickness, of surface orientation, of the type of the material, of a P_2O_5 treatment and of a wet-nitrogen treatment all become much less pronounced.

TABLE 7-III

Results of oxidation and treatments under conditions similar to those of table 7-I, but with little sodium contamination. Values of N_f are given for ⟨111⟩ and ⟨100⟩ surface-oriented samples

material	oxidation period (h)	after oxidation (O_2 1200 °C)					N_f ($cm^{-2} \times 10^{11}$) after various treatments					
		oxide thickness (μm)	Na in SiO_2		N_f ($cm^{-2} \times 10^{11}$)		N_2-H_2O 30 m 450 °C		P_2O_5 30 m 1050 °C		P_2O_5(1050 °C) + N_2-H_2O (450 °C)	
			top 100 Å (cm^{-2})	rest (average-cm^3)	⟨111⟩	⟨100⟩	⟨111⟩	⟨100⟩	⟨111⟩	⟨100⟩	⟨111⟩	⟨100⟩
5-Ωcm	¼	0·1	<2.10¹²	< 10¹⁷	5	2	4·5	2	5	0·5	5	1·5
p-Si	1	0·2	<2.10¹²	<5.10¹⁶	4·5	2·5	3	1·5	3·5	1·5	3	1·5
(B-dope)	4	0·4	< 10¹²	<2.10¹⁶	3·5	1	2·5	<0·5	2·5	<0·5	1·5	0·5
	16	0·8	<2.10¹²	< 10¹⁶	3·5	2	1·5	0·5	1·5	0·5	0·5	0·5
5-Ωcm	¼	0·1			4	0·5	3·5	1	4·5	0·5	2	2
n-Si	1	0·2	<2.10¹²	<5.10¹⁶	1	<0·5	1	<0·5	1	<0·5	1	0·5
(P-dope)	4	0·4			1·5	<0·5	0·5	1	1	<0·5	0·5	<0·5
	16	0·8			3	<0·5	1	<0·5	1	<0·5	0·5	<0·5

7.3.3. *Summary of results*

In a next section an attempt will be made to explain the various experimental results in terms of a physical-chemical model of an oxidized silicon surface. We shall now review the observed phenomena, including some results of earlier work [7-12,13]).

(1) In conventional silicon-oxidizing techniques contamination of the Si-SiO_2 system by sodium may occur during oxidation.
(2) The sodium concentration is then highest at the top of the oxide film, but there is also accumulation in the oxide and the silicon near the interface.
(3) The concentration of sodium at the Si-SiO_2 boundary becomes lower when the oxidation rate decreases, i.e. when the oxide has grown thicker.
(4) In the presence of sodium contamination the surface charge (N_f) decreases when the oxidation rate decreases.
(5) The value of N_f depends on the surface orientation of the silicon crystal, but most markedly when sodium is present.
(6) The value of N_f tends to be lower for *n*-type than for *p*-type material, especially in the case of ⟨111⟩ surface orientation and of sodium being present.
(7) A P_2O_5 diffusion (especially when carried out in a non-oxidizing gas) tends to reduce the amount of surface charge. The phosphate glass tends to getter sodium.
(8) The action of water vapour on the surface charge depends strongly on the temperature of heating. Below about 600 °C a considerable decrease in N_f values may be noted. At more elevated temperatures, however, such treatments may cause an increase [7-11,14]). Hydrogen acts in a similar way as water vapour, but the residual surface charge tends to be larger, especially at temperatures above 500 °C [7-13]). The hydrogen may also be evolved by a reaction between aluminium on the oxide and surface hydroxyl [7-12]).
(9) After oxidation or heat treatment in a dry environment, particularly when combined with a P_2O_5-diffusion treatment, large numbers of interface states can be present [7-10-16]). These may be distinguished as acceptor-type states close to the conduction band and donor states near the valence band [7-11]). They are probably due to the presence of unsaturated silicon bonds at the Si-SiO_2 interface.
(10) The number of interface states can be reduced drastically by low-temperature heat treatments in the presence of hydrogen or water [7-12,13]), i.e. the same treatment which can cause a decrease in N_f (note (8)). The effect of water vapour is largest when the water can be assumed to be reduced readily in the oxide film (when an oxide film has been heated in an inert gas after oxidation, even traces of water may be sufficient).

7.4. Discussion

7.4.1. *Interface states and the "flat-band" surface charge*

Before we shall discuss a model of an oxidized silicon surface, we will consider what kind of information is really obtained from the measurement of the "flat-band" point in a MOS C-V curve. Note (10) of the previous section suggests that the value of N_f is closely related to the interface states indicated in note (9). The major part of these interface states can be localized in regions from 0·1 to 0·3 eV from the silicon valence and conduction band [7-10,11]). In flat-band conditions on p-type material some states may be positively charged, but on n-type material the flat-band surface charge may tend to be negative. That such situations can indeed occur was shown by Whelan [7-15]) from C-V measurements at frequencies at 20 and 100 Mc/s (see fig. 2.23). However, at a frequency of 500 kc/s as used in the present work, the interface states can still follow the a.c. signal. When this occurs, the measured MOS capacitance is not in accordance with the band bending at the given d.c. voltage. Instead the oxide capacitance may be measured, as the states are situated at the oxide-silicon interface. This may occur over a wide range of the applied d.c. voltage. In the middle of the energy gap there are generally not a sufficiently large number of interface states to cause the exchange during the a.c. signal to occur completely at the interface. Consequently, also in the presence of a large number of interface states near the valence and conduction bands, the capacitance decreases below the oxide capacitance when the d.c. voltage causes the surface to be nearly intrinsic. The slope of the C-V curve looks then fairly ideal, i.e. as if there were no interface states present (fig. 2.23). There may be a displacement along the voltage axis, from which a number of charges N_f can be calculated. These N_f values may thus not always correspond to real flat-band conditions, but instead to a situation in which the charge in the donor-type interface states near the valence band and the acceptor-type interface states near the conduction band is approximately zero. This consideration leads to the conclusion that the N_f values determined from a 500-kc/s C-V curve do not give an indication of the number of these interface states. Instead, N_f should be related to a number of other states which contain a more or less fixed (positive) charge. One is inclined to ascribe these fixed charges to centres in the oxide structure. However, as was stated before, it is not possible from the present experimental information to distinguish between oxide states and donor-type interface states close to or above the conduction-band edge of the silicon.

More evidence that the surface charge corresponding to V_f is present in other states than those indicated in note (9) of the previous section, is obtained from the observation [7-11,13]) that a P_2O_5 diffusion causes in general an increase of the number of interface states but a decrease of the "flat-band" voltage determined from 500-kc/s C-V curves.

In a model for an oxidized silicon surface the centres which carry the surface charge N_f should thus be different from those causing the interface states near valence and conduction band. On the other hand a chemical relationship should be present (sec. 7.3.3, note (10)). The considerations given above indicate also that the differences noted between N_f values on n-type and on p-type samples (sec. 7.3.3, note (6)) cannot completely be explained by a difference in the occupation of the indicated interface states. There is probably also a difference in the number of other donor-type states, especially when impurities like sodium are present.

7.4.2. *Unsaturated silicon bonds at the* Si-SiO$_2$ *interface*

The valence of four which silicons shows can be related to the fact that a silicon atom has four electrons in its outer shell. When such an atom is bonded to less than four other atoms, the remaining unpaired electrons may be considered as unsaturated. They may be present as "dangling" bonds or give rise to double bonds between two atoms. In the following model we will try to consider unsaturated silicon bonds as a reason for various surface phenomena. For simplicity we will consider them as dangling bonds, of which we may distinguish various types, depending on their location in the Si-SiO$_2$ structure.

A dangling bond in the silicon lattice, i.e. an unpaired electron at a silicon atom bonded to three other silicon atoms, further indicated as Si$^{(3)}$, may be considered as being able to act as an acceptor

$$\begin{array}{c} \mathrm{Si} : \dot{\mathrm{Si}} : \mathrm{Si} \\ \ddot{\mathrm{Si}} \end{array} + e^- \rightleftarrows \begin{array}{c} \mathrm{Si} : \ddot{\mathrm{Si}} : \mathrm{Si} \\ \ddot{\mathrm{Si}} \end{array}$$

or

$$\mathrm{Si}^{(3)} + e^- \rightleftarrows \mathrm{Si}^{(3)-}. \tag{7.1}$$

However, the Si$^{(3)}$ centre may also be able to give off the unpaired electron and act as a donor:

$$\mathrm{Si}^{(3)} \rightleftarrows \mathrm{Si}^{(3)+} + e^-. \tag{7.2}$$

These types of centres may be assumed to occur at a clean silicon surface, but also when it is covered by SiO$_2$. Because of some misfit or lack of oxygen, not all silicon bonds may be saturated. The Si$^{(3)}$ centres may thus be considered as the reason for the (acceptor-type) surface states near the conduction band and the (donor) states near the valence band. During heat treatment in the presence of hydrogen the acceptor and donor action may disappear due to the reaction

$$\mathrm{Si}^{(3)} + \tfrac{1}{2}\,\mathrm{H}_2 \rightleftarrows \mathrm{Si}^{(3)}\text{-H}; \tag{7.3}$$

SiH bonds are probably less stable at elevated temperature and this explains why especially low-temperature heat treatments in the presence of H_2 or H_2O are able to cause the disappearance of interface states.

Apart from Si$^{(3)}$ centres, the Si-SiO$_2$ interface may also contain unsaturated silicon bonds at silicon atoms, which are bonded already to one, two or three oxygen atoms, and to two, one or zero silicon atoms, respectively. These may be indicated as Si$^{(2)}$, Si$^{(1)}$ and Si$^{(0)}$ centres. The Si$^{(0)}$ centre may also be considered as a half oxygen vacancy or as an Si^{3+} ion when the oxide structure is considered to be ionic. Summarizing:

$Si^{(0)}$: $$\begin{array}{c} O : \dot{Si} : O \\ \ddot{O} \end{array} \quad \text{or} \quad \begin{array}{c} O^{2-}\ Si^{3+}\ O^{2-} \\ O^{2-} \end{array}$$

$Si^{(1)}$: $$\begin{array}{c} O : \dot{Si} : O \\ \ddot{Si} \end{array}$$

$Si^{(2)}$: $$\begin{array}{c} Si : \dot{Si} : O \\ \ddot{Si} \end{array}$$

$Si^{(3)}$: $$\begin{array}{c} Si : \dot{Si} : Si \\ \ddot{Si} \end{array}$$

Due to the electronegative character of the surrounding oxygen ions the $Si^{(0)}$ centre can be supposed to act more readily as a donor than as an acceptor:

$$Si^{(0)} \rightleftarrows Si^{(0)+} + e^- \tag{7.4}$$

or

$$\begin{array}{c} O^{2-}Si^{3+}O^{2-} \\ O^{2-} \end{array} \rightleftarrows \begin{array}{c} O^{2-}\ Si^{4+}\ O^{2-} \\ O^{2-} \end{array} + e^-. \tag{7.5}$$

Also the $Si^{(2)}$ and $Si^{(1)}$ centres probably have more tendency to a donor action than to an acceptor action, compared to the $Si^{(3)}$ centre.

The surface charge measured from the flat-band point in a (500-kc/s) *C-V* curve must be related to centres with donor action, a role which might thus be fulfilled by the $Si^{(0)}$, $Si^{(1)}$ and $Si^{(2)}$ centres. The $Si^{(0)}$ centres may be present as well in the oxide as at the Si-SiO_2 interface, whereas one is inclined to consider the $Si^{(1)}$ and $Si^{(2)}$ centres to occur only at the interface. The donor action of the centres is in all cases related to the fact that they represent an incomplete oxidation product of silicon. We will further consider the $Si^{(0)}$ centres as an example, but one should thus realize that the $Si^{(1)}$ and $Si^{(2)}$ centres may behave similarly. A possible relationship between the presence of $Si^{(0)+}$ centres and impurities will be discussed in the next sections.

Until so far we have considered the $Si^{(3)}$ centres to cause both the presence of the acceptor- and the donor-type interface states near the conduction and valence band, respectively. It is not impossible, however, that at least a part of the donor states is due to $Si^{(2)}$ or $Si^{(1)}$ centres. All centres may react with hydrogen at low temperature, although some difference in reaction velocity or reaction equilibrium might be expected. This might explain why in certain hydrogen treatments the donor states near the valence band disappear less easily than the acceptor states near the conduction band [7–16]).

7.4.3. *The role of hydrogen in the structure of silicon dioxide*

The structure of SiO_2 may be considered as a network in which Si^{4+} ions are connected via bridging O^{2-} ions. Non-bridging O^{2-} ions may also be present when they are accompanied by "network-modifier" cations, such as protons and alkali ions. Such combinations can be assumed to be present due to incorporation of H_2O and Na_2O and so on in the SiO_2. The combination of a non-bridging oxygen ion and a proton is often called a hydroxyl group. It is well known, however, that hydroxyl groups can also form in SiO_2 as a result of heating in hydrogen [7-17,18]. This reaction has been described as

$$Si^{4+}\ O^{2-}\ Si^{4+} + \tfrac{1}{2} H_2 \rightleftarrows Si^{4+}\ O^{2-}\ H^+ + Si^{3+}, \tag{7.6}$$

although also the following possibility was considered [7-18]:

$$Si^{4+}\ O^{2-}\ Si^{4+} + H_2 \rightleftarrows Si^{4+}\ O^{2-}\ H^+ + H\text{-}Si^{3+}. \tag{7.7}$$

In the previous section the Si^{3+} ion ($Si^{(0)}$ centre) was assumed to be able to act as a donor at the silicon surface (reactions (7.5) or (7.4)). A reaction like (7.6) may thus be the reason for the formation of surface charge during heat treatment in hydrogen at elevated temperature (at too high temperatures — above about 1000 °C — the reducing action of H_2 may even cause the formation of volatile SiO). It was suggested before that the SiH bonds become less stable at elevated temperature, but reaction (7.7) cannot be excluded and may limit the number of surface donors. At low temperatures the formation of Si^{3+}-H (or $Si^{(0)}$-H) seems even very probable (reaction (7.3)).

7.4.4. *Oxidation of silicon in water vapour*

Oxidation of silicon in water vapour can be assumed to result also in the formation of hydrogen at the Si-SiO_2 interface and one may therefore expect reaction (7.6) to occur or, better, the oxidation is not likely to be completed directly at the interface due to a certain virtual hydrogen pressure at that place. Indeed it has been found that the efficiency of oxidation in steam can be much less than 100 per cent [7-19], while infrared measurements have been reported [7-20] which indicate a decreasing concentration of Si-O bonds in the oxide from the top towards the silicon surface. The concentration of hydrogen in steam-grown oxide films has been shown to be in the order of 10^{20} cm^{-3}, [7-28]. In the top layer of the oxide the hydrogen may be present due to the incorporation of water or OH groups. In the region near the interface the water tends to be reduced and OH groups combined with Si^{3+} ions or SiH groups may be present. Due to the donor action of the Si^{3+} ion ($Si^{(0)}$ centre) a positive oxide charge may tend to be formed, with the compensating electron occurring in the silicon:

$$Si^{4+}\ O^{2-}\ H^+\ Si^{3+} \rightleftarrows Si^{4+}\ O^{2-}\ H^+\ Si^{4+} \quad + e^-\ (Si) \tag{7.8}$$

or

$$Si - O - H\ \ Si^{(0)} \rightleftarrows Si - O - H\ \ Si^{(0)+} + e^-\ (Si). \tag{7.9}$$

The presence of these positive centres may also be described as an incomplete oxidation of silicon, which would be completed by reduction of the proton:

$$Si^{4+}\ O^{2-}\ H^+\ Si^{4+} + e^-\ (Si) \rightleftarrows Si^{4+}\ O^{2-}\ Si^{4+} + \tfrac{1}{2} H_2. \tag{7.10}$$

Oxidation of silicon in steam may thus result in the formation of positive oxide charge due to the formation of hydrogen at the Si-SiO_2 interface and its reducing action on SiO_2. The charge may also, however — although this means essentially the same — be related to incomplete oxidation of silicon or incomplete reduction of the water species. Whether one wants to ascribe the positive surface charge to the non-reduced protons or to the donor action of the $Si^{(0)}$ centres (compare the reaction products of reactions (7.8) and (7.9)), does not make any essential differ ence. The reduction of the positive surface charge by treatment in hydrogen at a relatively low temperature may be ascribed to the formation of $Si^{(0)}$-H bonds. The donor action of the $Si^{(0)}$ centres is then lost. One may also say that the positive charge of the non-reduced protons tends now to be compensated by the negative charge in the $Si^{(0)}H$ ($Si^{4+}H^-$) bonds instead of by electrons in the silicon. In the case of low-temperature treatment in the presence of water not only the formation of $Si^{(0)}$-H bonds but also the formation of $Si^{(0)}$-OH combinations may play some role in the reduction of surface charge.

7.4.5. *Oxidation of silicon in wet oxygen*

In this case there are two oxidizing species, oxygen and water. Hydrogen formed by the reduction of water at the Si-SiO_2 interface may move back towards the SiO_2-gas interface, until it meets neutral oxygen (somewhere in the oxide or at the oxide-gas interface) with which it reacts, after which oxygen ions together with protons move back to the silicon, and so on. In this way traces of water may act as a catalyst during oxidation in dry oxygen. Indeed it has been found that silicon oxidizes very slowly in extremely dry oxygen [7-21]). The virtual hydrogen pressure at the Si-SiO_2 interface will become lower when the oxygen content of the gas is increased, and this may explain why steam-grown oxides tend to induce more positive surface charge than (alkali-free) films grown in "dry" oxygen. It may be remarked that the values of N_f given in table 7-II for oxidation in wet oxygen cannot be ascribed completely to the role of hydrogen, as these films were not sodium-free.

7.4.6. *The role of sodium in the* Si-SiO_2 *system*

We have seen that during oxidation in "dry" oxygen the presence of hydrogen (water) may help the transport of oxygen ions through the oxide film. Apart from that, the presence of protons may catalyse the interface reaction due to the formation of intermediate structures such as shown in eqs (7.8) or (7.9). A similar role may be ascribed to sodium, which may be assumed to be present as sodium oxide in the oxidizing gas. Sodium ions moving together with oxygen ions towards the Si-SiO_2 interface tend to be reduced there. "Neutral" sodium tends to diffuse into the silicon and back in the direction of the oxide-gas interface. This may be considered as the reason for the peak in the sodium concentration at the Si-SiO_2 interface shown in fig. 7.1. Although neutral sodium in

SiO_2 probably does not exist, the virtual sodium pressure may give rise to the formation of associates of Na^+ and Si^{3+} ions: Si^{4+} O^{2-} Na^+ Si^{3+} or, in covalent notation, Si-O-Na $Si^{(0)}$. The motion of neutral sodium in SiO_2 may be considered as displacements of these centres due to motion of Na^+ ions together with unpaired electrons at the $Si^{(0)}$ centres. These electrons may also be given off to the silicon, so that donor action occurs. The ionized donors Si^{4+} O^{2-} Na^+ Si^{4+} or Si-O-Na $Si^{(0)+}$ may also be considered as intermediate reaction products. The density of these centres depends on the virtual sodium pressure at the interface. This pressure is determined mainly by the transport of Na^+ ions towards the Si-SiO_2 interface and the motion of "neutral" sodium into the silicon or back towards the oxide-gas interface, and thus by the oxidation rate and the sodium contamination in the oxidizing gas. This explains why the concentration of sodium at the Si-SiO_2 interface and the oxide charge tends to decrease with increasing oxide thicknesses. Table 7-I shows that the effect of oxide thickness remains present after a P_2O_5 diffusion. This is probably due to the fact that some oxidation of silicon still occurs during the diffusion (P_2O_5 may act as a source of oxygen). This would then mean that there is also some transport of Na^+ ions towards the Si-SiO_2 interface. In other experiments we have observed that the N_f values become higher when the P_2O_5 diffusion is carried out in oxygen instead of in an inert gas.

Not only does the virtual pressure of sodium at the Si-SiO_2 interface determine the surface charge. There is also an apparent influence of the interface structure, as is indicated by the different values of N_f found after oxidation of samples with different surface orientation. It is remarkable that the differences are most pronounced when sodium is present (compare tables 7-I and 7-III). Considering the formation of positive surface charge again as a result of a reducing action of sodium on the SiO_2 structure, it is plausible to assume that this occurs most readily in the interface region with the largest misfit between the structures of the silicon and the oxide. From these points of view the results suggest that the misfit is larger for $\langle 111 \rangle$ than for $\langle 100 \rangle$ surfaces. This is also consistent with the observation that the number of interface states, which in this work are attributed mainly to $Si^{(3)}$ centres, is largest for $\langle 111 \rangle$ surfaces [7-10,11]).

As was stated earlier [7-14]), the explanation for the effect of surface orientation may also be described as follows. Structures like Si-O-Na $Si^{(0)+}$ (Si^{4+} O^{2-} Na^+ Si^{4+}) represent an incomplete oxidation product of silicon. Transformation of these structures to Si-O-Si means a decrease of positive (oxide) charge (comparable to reaction (7.10)):

$$\text{Si-O-Na Si}^{(0)+} + e^-(\text{Si}) \rightleftarrows \text{Si-O-Si} + \text{Na}. \tag{7.11}$$

For this transformation the two silicon sites should be at a suitable distance, and the number of $Si^{(0)+}$ centres associated with the presence of sodium may thus be affected by the misfit between the oxide and the silicon structure.

In the presence of sodium there is also a clear influence of the type of material. The lower values of N_f for n-type instead of p-type material were also found by Revesz and Zaininger [7-6]), who suggested that the difference might be explained by incorporation of phosphorus in the interface structure. Considering the positive charge as being present in the oxide, it is more likely that the larger surface charge in p-type samples is due to incorporation of boron in the oxide structure, resulting in some way in an increased number of donor-type centres when sodium is also present. The difference between n- and p-type samples might also be explained by a difference in the frozen-in distribution of electrons between oxide and silicon due to a difference in electron concentration at the silicon surface during cooling. To eliminate this effect we did experiments in which the samples were exposed to light during heat treatment and cooling. Although some influence was noted, the major part of the difference of N_f in n-type and p-type samples remained.

It appears from table 7-II that alkali impurities are more effective in creating positive surface charge then hydrogen. This may be due partly to the possibility of formation of covalent bonds between $Si^{(0)}$ centres and hydrogen. Formation of $Si^{(0)}$-H bonds and perhaps some $Si^{(0)}$-OH bonds can also explain that the charge due to sodium can be decreased by heat treatment in water vapour and hydrogen. This can be understood to occur most effectively at low temperatures because of the instability of SiH bonds at elevated temperatures. Some exchange of sodium ions by protons may also occur and in this way the sodium concen-trations may be lowered (particularly during heating at high temperatures, table 7-II).

During various treatments there may also occur redistributions of charge in the oxide, which can cause changes in the C-V curves. These effects and also the influence of the way of cooling have been neglected in the given model*). There are certainly still other effects and impurities than those mentioned which can have some effect upon the surface properties of oxidized silicon. As an example we may remind of the fact that even temperature gradients during heat treatment can affect the charge distribution in an oxidized system [7-22]).

7.4.7. *Instability effects due to sodium-ion drift*

In oxide-coated silicon devices ion-drift effects may cause instability. This is in particular the case in devices containing MOS structures, where high electric

*) The experience that the amount of surface charge is increased when the oxidation temperature is decreased [7-6,11,29]) may be explained by high activation energies of reactions such as (7.10) and (7.11). Further, the diffusion of the oxidizing species into the oxide film will become more difficult when the temperature is decreased, whereas out-diffusion of hydrogen towards the oxygen-rich outer side of the film may still be relatively easy. A larger number of unsaturated silicon bonds near the oxide-silicon interface may then not be saturated with hydrogen. Fast cooling after oxidation may therefore yield lower values of N_f than slow cooling. However, after heat treatment in an inert gas, slow cooling may be preferred because traces of water (always present) may then cause a decrease of the number of positive surface charges and interface states (presence of water in an oxidizing gas is less effective in this respect; cf. table 4-I).

fields may be applied across the oxide film. It has been shown [7-15]) that the surface charge in oxide films prepared in a wet ambient can be fairly stable during heat treatments with a negative bias on the metal electrode of a MOS structure. This indicates that the protons present in the oxide film are not displaced easily. Under the same conditions somewhat larger changes of the oxide charge were found when the oxide was prepared in dry oxygen. This is probably due to displacement of Na^+ ions. However, it has not proved possible to make the surface charge zero or negative with such a treatment, except when the oxide contains trivalent ions (for example B^{3+} or Al^{3+}) [7-23,24]). From literature about vitreous systems it is known that the motion of univalent ions is then much easier [7-25]). In general, however, MOS structures show little instability during bias heat treatment with a negative bias on the metal electrode, which indicates that the sodium ions in the oxide films are not displaced very readily either.

Serious instability effects may be noted when the metal electrode is made positive during heat treatment. This is probably related to the drift of sodium ions, a considerable number of which may be present at the top of the oxide films. The drift effects are enhanced when some moisture is present [7-26]), suggesting that instabilities due to proton drift may occur too. However, it seems also possible that water decreases the activation energy necessary to make the sodium ions free to move into the oxide film. Instability effects can be limited by excluding contamination from alkali ions during the processing or by gettering them with the help of a phosphate glass. In the next section we will discuss an alternative method of removing alkali ions from oxide films.

7.5. Reducing treatments of oxidized silicon

Apart from at the Si-SiO_2 interface, the hydrogen and sodium concentration in the oxide may be related to an excess of oxygen being present. Kats [7-27]) has shown that quartz can be freed from hydroxyl (according to infrared measurements) by heat treatment in an inert gas or, more effectively, in a reducing environment of carbon monoxide. The same effect can be expected to occur in the case of sodium, so that such a treatment might offer the possibility of lowering the sodium concentrations in oxide films on silicon. Heat treatment in the presence of carbon monoxide or in high vacuum has the disadvantage that the oxide film tends to disappear due to the formation of volatile SiO. We have done experiments under reducing conditions in which this evaporation was avoided by heating the oxidized slices in an evacuated quartz capsule together with silicon powder. During heating a gas mixture forms, consisting mainly of SiO, formed by reaction between the silicon powder and the quartz wall, and further a very low concentration of silicon and oxygen. In such an environment the oxygen content of the oxide tends to become homogeneous throughout the film (at both sides excess oxygen is extracted by the silicon) and under these conditions Na_2O and H_2O tend to be reduced to Na and H_2, respectively. The sodium is then redistributed between the various phases present. Oxide films of 0·2 micron thickness made by oxidation of ⟨111⟩surface-oriented *p*-type silicon at 1200 °C in oxygen, showed a decrease in N_f from 15.10^{11} before to $1\text{-}3.10^{11}$ cm^{-2} after a treatment under such reducing conditions at 1050 °C for 30 minutes. Neutron-activation analysis showed a decrease of the average sodium concentration in the oxide by about a factor of ten, whereas diffusion into the silicon proved to have much increased. The low values of N_f would not be expected to occur if the sodium pressure in the system were high. The same may be said for the hydrogen pressure. We have indeed observed that N_f did not become very low when the system was not made sufficiently clean and dry. A low hydrogen pressure, probably always present, is favourable to the decrease of the number of unsaturated silicon bonds due to formation of SiH bonds. A disadvantage of the method is that the oxides may become leaky at weak spots.

During bias tests carried out at 150 °C with an electric field in the oxide of 10^6 V/cm, MOS structures made of oxide films with low sodium content showed no or little instability phenomena, and therefore it is apparent that alkali ions are the main reason for ion-drift phenomena in oxidized silicon.

7.6. Conclusions

Positive surface charge in oxidized silicon can be caused by the presence of a virtual sodium and hydrogen pressure at the Si-SiO_2 interface during heat treatment. These pressures are related to the amount of sodium and hydrogen (water) in the oxidizing gas and the oxidation rate of the silicon. Especially at low temperatures hydrogen and water tend to reduce the surface charge. These effects show a similarity to the disappearance of interface states during such treatments. Both low-temperature effects can be explained by the formation of SiH bonds.

The sodium concentration in thermally grown oxide films can be diminished by heat treatment under conditions of low oxygen pressure. Such treatments can be used to decrease the flat-band surface charge and to increase the stability of MOS structures.

REFERENCES

7-1) E. H. Snow, A. S. Grove, B. E. Deal and C. T. Sah, J. appl. Phys. **36**, 1664-1673, 1965.

7-2) E. Yon, W. H. Ko and A. B. Kuper, IEEE Trans. **ED-13**, 276-280, 1966.

7-3) H. G. Carlson, G. A. Brown, C. R. Fuller and J. Osborne, Paper presented at the Fourth Annual Physics of Failure in Electronics Conference at Chicago, Ill., III, November 1965.

7-4) A. S. Grove, B. E. Deal, E. H. Snow and C. T. Sah, Solid State Electronics **8**, 145-163, 1965.

7-5) J. R. Matthews, W. A. Griffins and K. H. Olson, J. electrochem. Soc. **112**, 899-902, 1965.

7-6) A. G. Revesz and K. H. Zaininger, IEEE Trans. **ED-13**, 246-255, 1966.

7-7) J. F. Delord, D. G. H. Hoffman and G. Stringer, Bull. Am. phys. Soc. **II 4**, 546, 1965.

7-8) P. Balk, P. J. Burkhardt and L. V. Gregor, Proc. IEEE **53**, 2133-2134, 1965.

7-9) Y. Miura, Jap. J. appl. Phys. **4**, 958-961, 1965.

7-10) P. V. Gray and D. M. Brown, Appl. Phys. Letters **8**, 31-33, 1966.

7-11) M. V. Whelan, Philips Res. Repts **22**, 289-303, 1967.

7-12) E. Kooi, Philips Res. Repts **20**, 578-594, 1965 (chapter 4 of this book).

7-13) E. Kooi, IEEE Trans. **ED-13**, 238-245, 1966.

7-14) E. Kooi and M. V. Whelan, Appl. Phys. Letters **9**, 314-317, 1966.

7-15) M. V. Whelan, Philips Res. Repts **20**, 595-619, 1965 (see also fig. 2.23 of this book).

7-16) P. Balk, Spring Meeting 1965 of the Electrochemical Society, Electronics Division, Abstract 109.

7-17) R. W. Lee, Phys. Chem. Glasses **5**, 35-43, 1964.

7-18) T. Bell, G. Hetherington and K. H. Jack, Phys. Chem. Glasses **3**, 141-146, 1962.

7-19) B. E. Deal, J. electrochem. Soc. **110**, 527-533, 1963.

7-20) H. Edagawa, Y. Morita, S. Maekawa and Y. Inuishi, Jap. J. appl. Phys. **2**, 765-775, 1963.

7-21) P. Balk, Fall Meeting 1965 of the Electrochemical Society, Electronics Division, Abstract 111.

7-22) E. Kooi and M. M. J. Schuurmans, Philips Res. Repts **20**, 315-319, 1965.

7-23) D. P. Seraphim, A. E. Brennemann, F. M. d'Heurle and H. L. Friedman, IBM J. Res. Dev. **8**, 400-409, 1964.

7-24) D. de Nobel, private communication.

7-25) G. Hetherington, K. H. Jack and M. W. Ramsay, Phys. Chem. Glasses **6**, 6-15, 1965.

7-26) S. R. Hofstein, IEEE Trans. **13**, 222-237, 1966.

7-27) A. Kats, Philips Res. Repts **17**, 133-195, 1962.

7-28) T. E. Burgess and F. M. Fowkes, Spring Meeting 1966 of the Electrochemical Society, Electronics Division, Abstract 55.

7-29) B. E. Deal, M. Sklar, A. S. Grove and E. H. Snow, J. electrochem. Soc. **114**, 266-274, 1967.

List of symbols

C	differential capacitance per cm^2 of a MOS diode
C_{min}	the minimum capacitance value in the C-V curve of a MOS structure (per cm^2)
C_{ox}	capacitance of the oxide film per cm^2; $C_{ox} = \varepsilon_{ox}/d_{ox}$, where ε_{ox} ($=3{\cdot}4.10^{-13}$ F/cm) is the dielectric constant of the oxide and d_{ox} the oxide thickness in cm
C_{Si}	silicon space-charge capacitance per cm^2
C_{Si_0}	silicon space-charge capacitance for flat-band conditions per cm^2
$C_{Si\,min}$	the minimum silicon space-charge inversion capacitance per cm^2
e	the unit charge, $1{\cdot}6.10^{-19}$ coulomb
E_F	Fermi level
E_g	energy gap in a semiconductor (1·1 eV for Si at 300 °K)
E_i	intrinsic energy level (in the middle of the band gap)
E_s	band bending at the surface (in eV), $E_s = -eV_s$
F	see at V_0
g_m	transconductance of a MOS transistor, defined as $g_m = \left(\frac{\partial I_D}{\partial V_G}\right)_{V_D}$ or, in the region where the drain current saturates $g_m = \frac{d\,I_D(\text{sat})}{d\,V_G}$
I_D	the drain current of a MOS transistor flowing when a voltage is applied between drain and source
I_D(sat)	saturation value of the drain current
k	Boltzmann's constant
L	channel length of a MOS transistor (source-drain distance)
n	electron concentration (per cm^3)
n_i	electron and hole concentrations in intrinsic material (for silicon $n_i = 1{\cdot}6.10^{10}$ cm^{-3} at 300 °K)
N	number of charge carriers per cm^2 surface area
N_f	effective charge density (per cm^2) at the Si-SiO_2 interface, determined from the flat-band-capacitance point in a MOS C-V curve
p	hole concentration (per cm^3)
Q_{depl}	immobile (depletion) part of the space charge at the silicon surface (per cm^2)
Q_{inv}	mobile (inversion) part of the silicon space charge (per cm^2)
Q_M	charge on the metal electrode in a MOS structure (per cm^2)
Q_{ox}	effective oxide charge (per cm^2) represented as a sheet of charge at the Si-SiO_2 interface
Q_{Si}	space charge at the silicon surface (per cm^2)
Q_{ss}	charge in surface states (per cm^2) at the Si-SiO_2 interface

V_0 — a parameter in the relation between I_D(sat) and V_G of a MOS transistor; $I_D(\text{sat}) = \frac{1}{2} F\beta(V_G - V_0)^2$, in which $\beta = \varepsilon_{ox}\,\mu W/d_{ox}L$ and F a constant (between 0 and 1) depending on the doping level of the substrate material

V_D — voltage applied between drain and source of a MOS transistor

V_f — the voltage which has to be applied on the metal electrode of a MOS structure to get flat-band conditions in the silicon

V_G — the gate voltage (with respect to the source) in a MOS transistor

V_i — the voltage which has to be applied to the metal electrode of a MOS structure to get an intrinsic surface

V_M — the voltage of the metal electrode of a MOS structure with respect to the substrate

V_P — the drain voltage needed to get channel pinch-off in a MOS transistor

V_s — surface potential, related to the band bending by $E_s = -eV_s$

V_T — threshold voltage of the gate voltage, where a certain (to be defined) amount of drain current can start to flow

W — channel width of a MOS transistor

y_s — eV_s/kT

β — see at V_0

λ — n_0/n_i for n-type material, p_0/n_i for p-type material (n_0 and p_0 are the equilibrium electron and hole densities of the given material, $n_i = 1{\cdot}6.10^{-10}$ cm^{-3} at 300 °K)

μ — mobility of charge carriers